Future City

Volume 8

Future City Description

As of 2008, for the first time in human history, half of the world's population now live in cities. And with concerns about issues such as climate change, energy supply and environmental health receiving increasing political attention, interest in the sustainable development of our future cities has grown dramatically.

Yet despite a wealth of literature on green architecture, evidence-based design and sustainable planning, only a fraction of the current literature successfully integrates the necessary theory and practice from across the full range of relevant disciplines.

Springer's *Future City* series combines expertise from designers, and from natural and social scientists, to discuss the wide range of issues facing the architects, planners, developers and inhabitants of the world's future cities. Its aim is to encourage the integration of ecological theory into the aesthetic, social and practical realities of contemporary urban development.

More information about this series at http://www.springer.com/series/8178

Joe R. McBride

The World's Urban Forests

History, Composition, Design, Function and Management

Springer

Joe R. McBride
Landscape Architecture and Environmental
University of California Landscape
Architecture and Environmental
Berkeley, CA, USA

ISSN 1876-0899 ISSN 1876-0880 (electronic)
Future City
ISBN 978-3-319-84822-8 ISBN 978-3-319-52108-4 (eBook)
DOI 10.1007/978-3-319-52108-4

Softcover reprint of the hardcover 1st edition 2017

Printed on acid-free paper

This Springer imprint is published by Springer Nature
The registered company is Springer International Publishing AG
The registered company address is: Gewerbestrasse 11, 6330 Cham, Switzerland

Preface

Pliny the Elder, in the preface to his *Natural History*, described his book as a study of "the nature of things." My ambition, in this natural history of the world's urban forest, is more limited; I intend only to address the nature of trees occurring in urban areas. My purpose is to examine the history, composition, design, function, and management of urban forests in 33 of the world's cities. Of particular interest in this examination is the question of the relationship of urban forest history, composition, design, and management to the large geographic regions, known as biomes, in which these cities occur. An initial assumption to be tested is the degree to which the climates and arboreal floras of the biomes dictate the characteristics of local urban forests. History, no doubt, has also played a role in shaping the characteristics of present-day urban forests. Therefore, the book will also investigate the influence of culture and world history on the urban forest.

The meaning of the term *urban forest* varies throughout the world. In Germany, it refers to a legally designated forest near the margin of a city that is managed specifically for the use of the people living in the city. In Sweden, the term applies to forests in urban areas where the understory vegetation is not managed. In the United States, the term *urban forest* is used to encompass all trees growing in an urban area, regardless of how they are managed. This definition I found the most encompassing and useful in the study of trees in urban areas around the world.

The history of trees in urban areas is not limited to the 33 cities that were studied for this book. Many precedents for the use of trees in cities were developed in other cities and even outside of the boundaries of urban areas. For example, commonly used planting patterns, such as the allee, were developed in Renaissance gardens before they were adopted in urban areas. The historical analysis presented in this book depended primarily on library sources. These sources were augmented by examining paintings, maps, and other representations of trees in museums in many of the cities studied. In some cities, it was possible to visit historic boulevards and parks to experience historically significant designs that were innovations in their time.

Street and boulevard trees as well as trees in parks, plazas, gardens, cemeteries, greenbelts, and certain special urban environments (e.g., urban land adjacent to water edges, traffic circles, rooftop gardens, trees on walls) were examined to gather information for this book. The collected data are presented in tables illustrating the composition of urban forest in the 33 cities studied. Simple averages and calculations of frequency of occurrence are used to give a general picture of the importance of different species in various urban environments. A comparative analysis of planting designs, illustrated with cross-section diagrams and sketch maps, is used to contrast designs used in different cities.

Information on the management of the world's urban forests came from interviews in each city. Urban foresters, arborists, landscape architects, city planners, park managers, and local citizens were asked about issues concerning the management of trees. Although each city had its special management issues, some issues such as pest management and funding were common to most of the cities studied.

This book brings together an analysis of the world's urban forests from the perspective of the ecological influence of the biomes and cultural and historical factors influencing the development of cities. Strong correlations can be drawn between the characteristics of certain biomes and the composition, design, and management of urban forest in those biomes. But this relationship is not necessarily universal. Culture and historical development of cities have played more dominant roles in the development of urban forests in many of the world's biomes. The future of the world's urban forests is treated in a final chapter that identifies contrasting trends. Some of these trends will enhance and enrich urban forests, while others are leading to a decline in tree species diversity and canopy cover in many of the world's cities.

Preparing this book involved flying many thousands of miles to reach the dispersed cities, walking hundreds of miles of streets to survey urban forest conditions, conversing with dozens of local tree experts, and spending uncounted hours writing late into the night. The effort has rewarded me by broadening my perspective on the urban forest and giving me a greater appreciation for the variety of ways trees can be used in urban areas. And I believe, as did Pliny, those sleepless nights spent in writing have added hours to my life. He said in justification of the late hours spent in writing his *Natural History* – "to be awake is to be alive."

Berkeley, CA, USA Joe R. McBride

Acknowledgments

This book would not have been possible without the help of graduate students, colleagues, and my wife, Deborah McBride, who accompanied me to many of the cities included in this book. Their assistance in making travel arrangements, contacts with local urban forestry officials, and data collection and their skills, as translators of a variety of languages, were invaluable. In particular, I wish to thank the following individuals for their assistance in the cities indicated: Ahmad Mossadegh (University of Tehran), Tehran and Dubai; Zemin Wu (Anhui Agricultural University), Beijing; Jan Falck (Swedish University of Agricultural Sciences); Sheauchi Cheng (US Department of Agriculture), Tokyo; Vlad Douhovnikoff (Bowdoin College), Murmansk and Moscow; Scot Medbury (Brooklyn Botanic Garden), Moscow; Jun Yang (Beijing Forestry University), Beijing; Danai Thaitakoo (Chulalongkorn University), Bangkok; C. Y. Jim (University of Hong Kong), Hong Kong; and Trausti Valsson (University of Iceland), Reykjavik. Numerous local officials in urban forestry were also most helpful in providing information about trees and their management in the various cities. Their generosity in taking time to meet with me and answer my questions is most appreciated. I am also indebted to my friend Dick Berman who provided valuable counsel throughout the process of producing this book and specific assistance in the use of various types of computer software. Support for the research necessary for this book was provided by the Farrand Fund for Research in Landscape Architecture and the Pacific Rim Research Program of the University of California.

A special word of gratitude is due to Deborah McBride who accompanied to many of the cities and assisted in a variety of ways. Without her confidence in the value of the project and her encouragement, this work would not have been completed. Finally, this book would not have been possible without the continued encouragement of Jim Clark of HortScience.

Contents

Chapter 1
Introduction

Trees have occurred in urban areas since the evolution of the city, however, their importance has varied with time. Initially restricted to the gardens of the aristocracy, trees became common elements of urban streetscapes and public parks in the late Renaissance. During the nineteenth and twentieth centuries leafy suburbs arose around many cities in Europe and North America expanding urban tree cover and increasing species diversity. Today, trees continue to be important to the urban landscape. One can hardly reflect on Paris without the mental image of the tree-lined Champs Elysees, Washington, DC without the American elms along the Mall, or Singapore without its lovely angsana tree-lined avenues.

The rulers of early cities and their advisors were responsible for the selection of tree species for planting and for the design of private gardens and public spaces. Later in the evolution of the city, specialists in town planning, architecture, and landscape architecture were responsible for designing street tree plantings, public parks and gardens. The influence of common citizens on tree selection and design, outside of privately owned property, was of only sporadic importance until the nineteenth and twentieth centuries when citizen participation became an important part of planning in some of the world's cities.

During the evolution of the city, maintenance of trees has been the domain of arborists and gardeners who cared for trees in public open spaces, city parks and along streets. The role of traditional foresters in the planning and management of trees in urban areas was limited until the latter half of the twentieth century. Then individuals trained in forestry were brought into city agencies responsible for the planning and maintenance of urban trees and the term "urban forestry" gained widespread usage. Foresters introduced new ecological and economic concepts for the collective management of trees in urban areas. In North America urban foresters developed a rather inclusive view of the urban forest. In Europe, the tendency was to distinguish trees within a city from trees in greenbelts and forested parks adjacent to the city.

The purpose of this book is to examine urban forests in cities around the world. It will ask questions about the history, composition, structure, and management of

J.R. McBride, *The World's Urban Forests*, Future City 8,
DOI 10.1007/978-3-319-52108-4_1

trees in urban areas. Data for this book was collected in 33 cities across broad geographical areas known as biomes. Constraints and opportunities imposed on urban forest composition, design, and management by the ecological characteristics of these biomes will be examined. The book will also address the cultural and historical factors that influenced the characteristics of urban forests around the world.

The term "biome" is used to refer to a biotic community composed of plants and animals that share adaptations to environmental conditions in a broad geographical region. It is similar to the term ecosystem, but has been rejected by ecosystem scientists because of the use of vegetation as the defining characteristic of each biome (Bailey 1996). However, the specific use of vegetation as a defining characteristic makes the biome especially suited for this investigation. A principal assumption of this book is that the regional vegetation that defines each biome has had an important influence on the design of the urban forests within the biome. Tree species may be selected from the regional flora to compose the urban forest. The pattern, density, and vertical structure of the natural forests in the biome may also provide an important context for the design characteristics of the urban forest. In addition, the climatic conditions of the biome may limit the tree species that can be used in the local urban forest. Exploring the relationships between urban forests and the biomes in which they occur is a major objective of this book.

Several classification systems of biomes were reviewed before selecting one for this study (Finch and Trewartha 1949; Bartholmew 1950; Hammonds 1961; Eyre 1968; Breckle 2002). The differences among these classifications frequently had to do with the numbers of biomes identified and their geographical boundaries. For example, Finch and Trewartha (1949) suggest two categories of tropical forest (tropical rain forest and semi-deciduous tropical forest) based on seasonal rainfall patterns and leaf retention. These categories were merged into a single type (tropical rain forest) by Bartholomew (1950) on the basis of similarity of genera, high annual rainfall, and absence of freezing temperatures. Eleven widely agreed-upon biomes are used in this book. These are the tropical forest, broadleaf evergreen forest, deciduous forest, coniferous forest, mixed deciduous-coniferous forest, Mediterranean woodland/scrub, savanna/scrub, grassland, desert, tundra, and highlands biomes. The cities selected for study are listed in Table 1.1 according to the biomes in which they occur.

The ecology of these biomes has not determined the characteristics of the world's urban forests alone. Cultural and historical factors have also had their influence on the composition, structure, and management of urban forests. One may be able to identify the importance of cultural and historical factors by contrasting urban forests in cities that have emerged from different cultures or along different lines of historical development within the same biome. For example, the design of traditional Japanese gardens compared to Italian Renaissance gardens shows recognizable differences due to cultural influences. The design of gardens, parks, and boulevards in many cities show an historical influence of foreign colonial powers that imposed their cultural values upon existing cultures during periods of colonial domination. A review of the history of the use of trees in cities will provide a better

Table 1.1 Cities surveyed in the study of the world's urban forests

Biome	Cities
Tropical forest	Bangkok, Singapore, Rio de Janeiro
Broadleaf evergreen forest	Hongkong, Sydney, Tokyo
Deciduous forest	London, New York, Paris
Coniferous forest	Jacksonville, Seattle, Vancouver
Mixed deciduous-coniferous forest	Beijing, Moscow, Stockholm
Mediterranean woodland/scrub	Athens, Los Angeles, Tel Aviv
Savanna/scrub forest	Brasilia, New Delhi, Lagos
Grassland	Buenos Aires, Johannesburg, Tehran
Desert	Cairo, Dubai, Phoenix
Tundra	Murmansk, Nuuk, Reykjavik
Highland	Cusco, Lhasa, Quito

understanding of the variations observed in tree species composition and planting patterns in various urban areas.

The ecologic characteristics of a biome may result in specific management problems for urban forests. The lack of soil moisture in the desert biome demands the use of irrigation for the establishment and survival of trees not adapted to the desert environment. Native forests in tree-dominated biomes may serve as sources for the spread of insects or diseases that attack trees in urban areas. Such problems do not exist for urban forests in the grassland, desert, or tundra biomes. Management practices vary as a result of cultural differences as illustrated by tree pruning practices in different cities. The involvement of local citizens in the planning and maintenance of urban forest trees may be the result of the historical development of government policies in different cities. Identifying and understanding the sources of urban forest management practices will be a part of this book.

The book identifies special uses of trees observed in the cities studied. All cities use trees to line streets and boulevards and to enhance to quality of urban parks. In addition to these uses some cities have employed trees for special purposes. For example, Oriental plane trees are planted in artificial canals in Tehran to evoke the riparian forests which once crossed the great alluvial fan on which the city was built. This particular use of trees is at once a reminder of the ecological heritage of the city and a means of creating a pleasant urban microclimate. In Tokyo trees are elements of small parks designed as habitats for insects such as dragonflies and lightning bugs. Trees enhance the rooftop gardens in New York City and other cities to provide both shade and wind protection for rooftop users. In each of the cities studied an effort was made to identify examples of special uses of trees that expand the nature of the urban forests.

The future of the urban forest will be is discussed in the last chapter of the book. At the end of the twentieth century the number of trees and tree canopy cover were declining in many cities in the United States (Moll 1997). At the same time many expanding cities in Asia were increasing the number of trees and the amount of per capita green space. The introduction of European concepts of trees in urban design in African and South American colonies in the eighteenth and nineteenth centuries initiated urban forests that do not respond to the needs of many of the burgeoning populations of post-colonial cities on these continents. The degree to which urban forests have met their potential both in the developed and developing parts of the world can be seen by examining the functional roles urban forests play. Conclusions drawn from this study will suggest future trends in urban forestry and how these trends may need to be modified if urban forests of the twenty-first century are to continue to grace our cities and effectively serve the people living in them.

Chapter 2
Methods

Information was collected for this study by sampling specific characteristics of the urban forests in the 33 of the world's cities. An analysis of aerial photographs and maps was employed to determine the amount of urban tree cover. Literature relating to urban forest conditions and the history of trees in cities was reviewed. Interviews were conducted with city planners, landscape architects, arborists, urban foresters, park officials, and academics involved in the planning and management of the urban forests in the cities studied for this book. Local citizens were also questioned about the characteristics and values of trees in their cities. The data and information collected was analyzed to identify the characteristics of the urban forests and to understand the importance of the ecology of the biomes in which the cities occurred as well as the cultural and historical influences that shaped each urban forest.

Sampling of urban forests in the cities studied provided data on the tree characteristics that could be summarized according to the following urban environments:

1. Streets and boulevards
2. Parks
3. Plazas
4. Cemeteries
5. Water's edge
6. Waste areas
7. Special environments (e.g., traffic circles, roof top gardens, freeways, etc.)

The data collected included the species of trees, diameter at breast height, tree height, height to the base of the live crown, crown radius, growing space, tree spacing, canopy closure, and tree condition. This information was recorded on the form shown in Fig. 2.1.

Tree identification was made in each city using local taxonomic keys where local keys were available. Otherwise regional or international keys were used. Both local common names and scientific names are used in this book. The scientific names used in Krussmann's manuals of cultivated trees were used when differences in scientific names for the same species were encountered among different cities,

J.R. McBride, *The World's Urban Forests*, Future City 8,
DOI 10.1007/978-3-319-52108-4_2

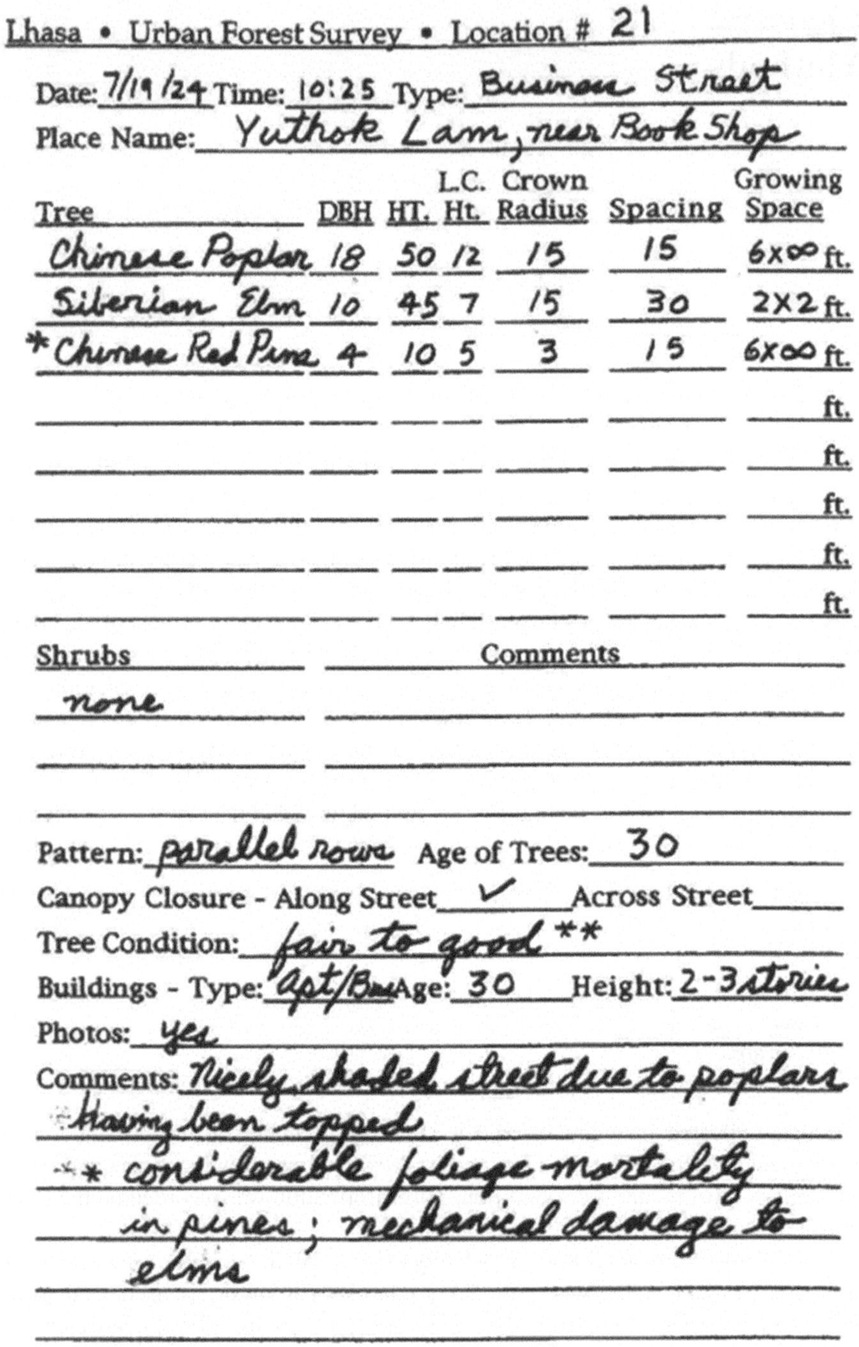

Lhasa • Urban Forest Survey • Location # 21

Date: 7/19/24 Time: 10:25 Type: Business Street

Place Name: Yuthok Lam, near Book Shop

Tree	DBH	HT.	L.C. Ht.	Crown Radius	Spacing	Growing Space
Chinese Poplar	18	50	12	15	15	6x∞ ft.
Siberian Elm	10	45	7	15	30	2X2 ft.
*Chinese Red Pine	4	10	5	3	15	6x∞ ft.
						ft.
						ft.
						ft.
						ft.
						ft.

Shrubs	Comments
none	

Pattern: parallel rows Age of Trees: 30

Canopy Closure - Along Street ✓ Across Street

Tree Condition: fair to good **

Buildings - Type: Apt/Bus Age: 30 Height: 2-3 stories

Photos: yes

Comments: Nicely shaded street due to poplars Having been topped

** considerable foliage mortality in pines; mechanical damage to elms

Fig. 2.1 Data form

(Krussmann 1984, 1985). Illustrations of the leaves of the more common species were drawn from specimens collected in various cities.

Sampling sites were initially located on a random basis by laying a grid over a map of a city and picking intersection points of the grid lines using a table of random numbers. In using this method it was found that a large percentage of the sample points selected at random did not support any street trees. The method failed to locate sample points in many of the urban environments of interest, such as cemeteries and small parks. This method was abandoned in favor of a procedure in which local urban forest experts and citizens were asked to suggest a number of streets and boulevards that represented the range of tree characteristics in their city. Recommendations of parks, cemeteries, and water's edge environments were also sought. Between 30 and 50 sites were sampled in each city, with a minimum of three sites surveyed in each of the major urban environments. In some cities additional sampling sites were selected to include historically significant examples of the urban forest. The final selection of sampling sites cannot be considered a random selection and, therefore, statistical comparisons of the data are not justified. However, the frequencies of occurrence of tree species observed in various urban environments were calculated. Comparisons of these frequencies of occurrence to published species rankings based on complete tree inventories in some of the cities suggest the method of sampling provides a accurate estimate of the more commonly used tree species.

Ocular estimates of tree dimensions and tree planting spaces were used after initial calibration of the eye with a diameter tape, clinometer, and measuring tapes (for crown radius and tree planting spaces). Tree spacing was determined by pacing, as were street and sidewalk widths. These too were checked periodically with a measuring tape.

In addition to the data on tree characteristics recorded on the tree data form, cross-section diagrams were made of the street and boulevard sample sites and site maps were drawn of small parks, cemeteries, plazas, and water edge environments. Examples of these are shown in Fig. 2.2 (street cross section), Fig. 2.3 (small park), and Fig. 2.4 (water's edge).

Interviews with city planners, landscape architects, arborists, urban foresters, park officials, and academics focused on issues of the design and management of the urban forest. Attempts were made to meet with the heads of municipal agencies responsible for urban forest management. This was not always possible because of time constraints. The following questions were asked of the individuals interviewed:

1. What agency (or agencies) is responsible for the planning and management of the urban forest?
2. What are the major urban forest management issues?
3. To what degree is the community involved in urban forest planning and management decisions?

These questions were intended not only to elicit specific information, but also to stimulate an open-ended discussion about the planning and management of the

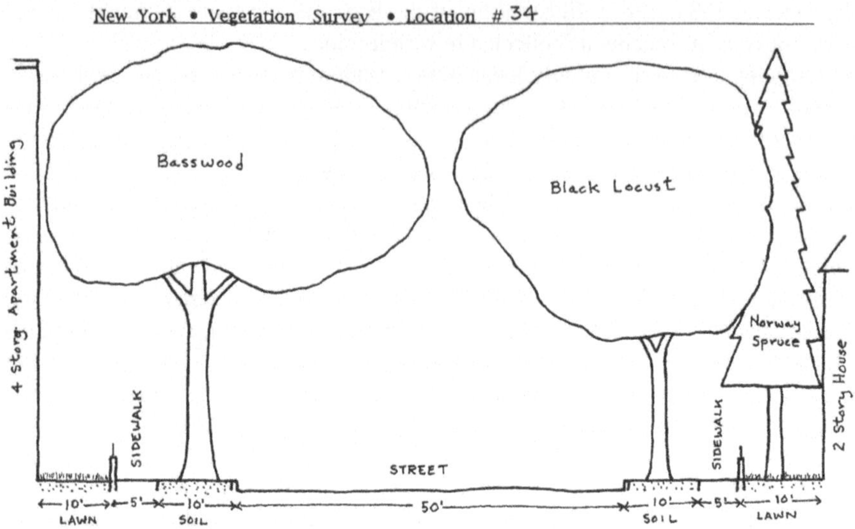

Fig. 2.2 Street cross-section

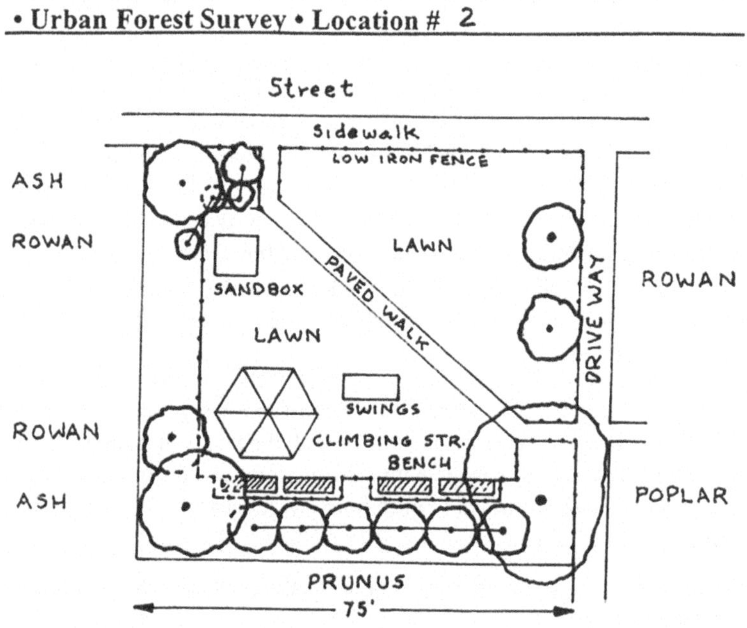

Fig. 2.3 Sketch map of small park

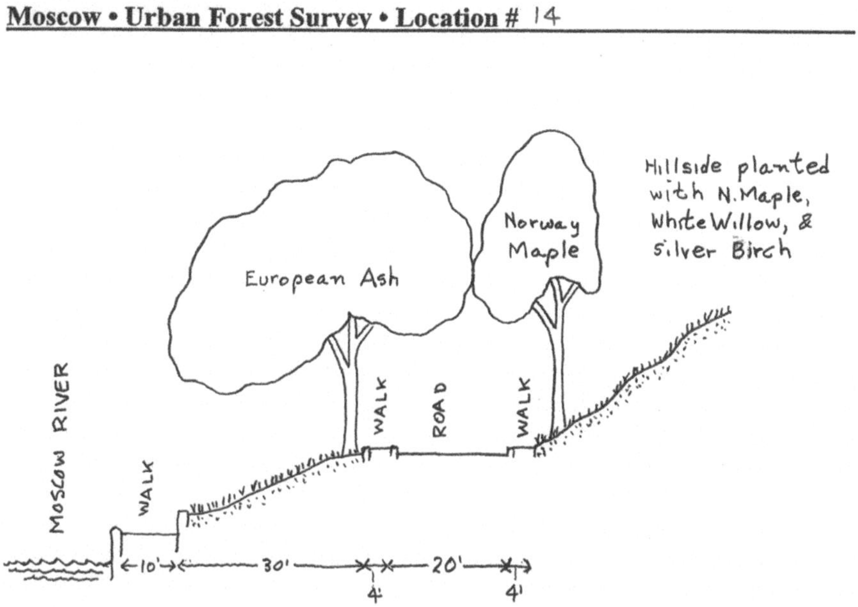

Fig. 2.4 Water's edge cross-section

urban forest. Interviews usually lasted 1 h. In some cases the initial individual interviewed arranged for members of his or her staff to be available for additional interviews or to provide a tour of facilities related to urban forest management (e.g., municipal tree nursery). Whenever possible, tree workers were approached on streets or in parks and asked about their work and views on local tree management. Cab drivers, hotel clerks, waiters, and people on the street or in parks were asked about the trees in their city. Note taking and drawing of cross-section diagrams often attracted the attention of people passing by. Conversations with these people about trees sometimes arose.

Chapter 3
Biomes, Human Settlement, and the Selection of Cities

Biomes

The natural world may be divided into large regional areas on the basis of vegetation and climate that are referred to as biomes. The names used for these natural regions are often derived from characteristics of the vegetation and climate, for example the "tropical rain forest". The tropical rain forest or any other of the world's biomes have common characteristics related to the climate and vegetation that allows one to recognize different examples on different continents. Although the tree species may vary between the forests of Borneo, the Amazon basin, and the Congo one can easily recognize the similarity of the structure of the forests and persistent rainfall of these areas. These worldwide natural areas are the products of a long evolution and migration of plants and animals in response to changing climatic gradients. These natural regions appear to be stable in the context of the human time scale, but are dynamic in the context of geologic time.

Within each of the world's large natural areas are many sub-regional communities of plants and animals interacting with one another and with the local environment. The German zoologist, Karl Mobius, recognized the interactive nature of these relationships and suggested the term "biocoenosis" to identify this relationship using as an example the complex system of oysters, the plankton they feed on, and the environment of the oyster beds in his classic study (Mobius 1877). In the twentieth century, other terms such as "biome" (Shelford 1932) and "ecosystem" (Tansley 1935) were proposed for communities of plants and animals and their environments. The most widely accepted of these new terms was the word "ecosystem", which was suggested as a term in reference to any community of plants and animals together with their abiotic environment, forming an interacting system. Tansley's use of the term was without reference to scale and could describe the oyster beds studies by Mobius or the large regional biomes defined by their vegetation as proposed by Shelford. In this book the term biome is used in reference to the world's large natural regions. The larger scale connotation of the term biome is most

J.R. McBride, *The World's Urban Forests*, Future City 8,
DOI 10.1007/978-3-319-52108-4_3

appropriate to the worldwide focus of this book. Its use should avoid any confusion with local ecosystems.

The use of the term biome puts the emphasis on vegetation and avoids complications of identifying ecosystems at a global scale (Bailey 1996). Bailey's concern with the use of vegetation to define ecosystems at a global scale was that vegetation within any region has a dynamic character due to natural succession and manipulations by humans, so that at any one time the present vegetation may not be indicative of the climax. Biomes are the appropriate focus for this book since they are based on large regional vegetation units that have been mapped by various geographers (Kuchler 1967, 1973) focusing on the potential climax vegetation of the region.

Historical Background

The geographical delineation of the world's biomes is based on the distribution of major vegetation types. Early geographers mapped the world's vegetation before the term biome had been proposed. By the middle of the nineteenth century enough information had been amassed to produce relatively accurate world maps of both vegetation and climate. A general relationship between the distribution of vegetation and climate had been understood since the time of the ancient Greek philosophers. Theophrastus (372–287 BC) recognized the importance of topographic position and climate in the distribution and growth of plants. He had contrasted the characteristic tree species of the Mediterranean region with those of the colder regions known to the Greeks. The distribution of vegetation and its relation to climate was further described by the Greek geographer Strabo (64–23 BC) and the Roman natural historian Pliny the Elder (77 AD).

Exploration of the western hemisphere during the Renaissance brought new information about world vegetation and climate back to Europe. A conceptual framework for understanding the accumulated information on vegetation and climate was advanced by the German explorer and scientist Alexander von Humboldt. In 1805 he published an essay on plant geography in which he identified the relationship between elevation, climate, and vegetation type distribution in the tropics (von Humboldt 1805). His remarkable drawing of the distribution of vegetation on the volcano Chimborazo, near the equator in Ecuador, illustrated the distribution of vegetation types in relation to altitude (Fig. 3.1). Subsequent geographers extrapolated von Humboldt's elevation gradient near the equator to the change in vegetation types along the latitudinal gradient from the equator to the poles (Dansereau 1957). These extrapolations identified the similarity of vegetation types as one moves up in elevation or latitude from tropical rain forests near the equator to the tundra vegetation of high elevations and polar latitudes (Fig. 3.2).

Von Humboldt's work inspired the Swiss botanist Augustin de Candolle to produce the first modern vegetation map of the world (de Candolle 1820). The vegetation classification used by de Candolle to map the world's vegetation was based on physiognomic types. These types were determined by the life form of the dominant

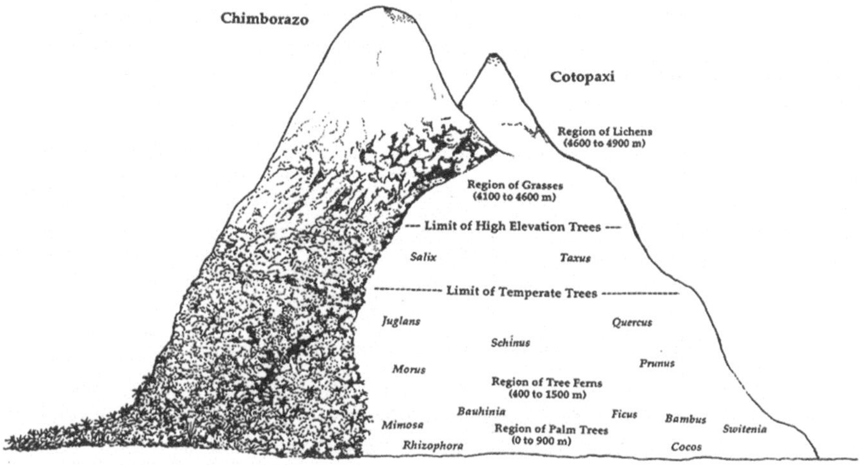

Fig. 3.1 Distribution of vegetation on Chimbarazo (Redrawn from von Humboldt 1805)

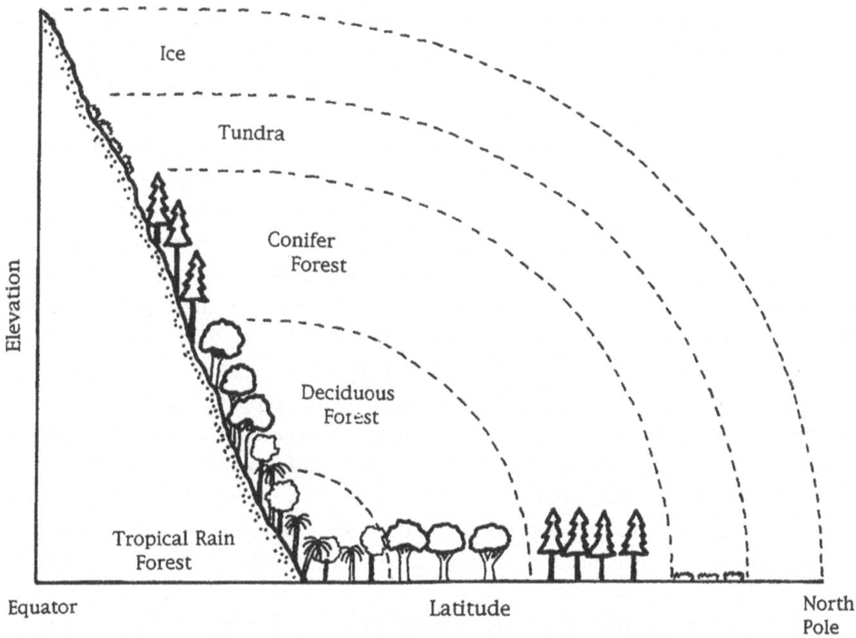

Fig. 3.2 Generalized distribution of vegetation in relation to elevation and latitude

plants, the mature heights of these plants, and their spacing. The principle physiognomic types are forest, woodland, savanna, scrub, grassland, and tundra (Fig. 3.3). This structural approach to vegetation classification did not require knowledge of the taxonomy of the regional flora. It was relatively easy to apply because of the

Type	Forest	Woodland	Savanna	Scrub	Grassland	Tundra
Section						
Life form	Tree	Tree	Tree/Grass	Shrub	Grass	Lichen/Grass
Height (ft)	> 30	< 30	varied	< 6	< 3	< 1
Spacing	canopies touching	canopies touching	canopies not touching	canopies usually touching	not applicable	not applicable

Fig. 3.3 Characteristics of major physiognomic vegetation types

distinctive nature of the different physiognomic types. As an approach to vegetation classification and mapping, it was found to be very adaptable to remote sensing techniques that developed in the twentieth century (Ellenberg and Mueller-Dombois 1973).

Von Humboldt's interest in climate stimulated other scientists to address the question of climate classification and mapping. Following his return from South America, von Humboldt established a series of weather station across Europe and into Siberia. The data collected at these stations along with data from locations around the world allowed the Russian climatologist Wladimir Köppen to produce a world map of temperature belts in 1874 and his mathematical system of climate classification in 1900. In attempting to find significant temperature boundaries for his climatic types, Köppen often relied on vegetation boundaries drawn by de Candolle. For example, Köppen set the temperature limit of his tropical rainy climate at an average temperature every month above 18°C [64.4°F] because the boundary established by this temperature corresponded with the boundary of the tropical rain forest mapped by de Candolle (Köppen 1931). In the twentieth century various plant geographers, climatologists, and ecologists worked to refine the classifications and maps developed by de Candolle and Köppen. These refinements incorporated new information on the distribution of vegetation types, new climatic data, and new ideas about the nature of vegetation and climate. Among these new ideas was the vegetation classification system based on the location of the renewal bud of a plant proposed by Raunkiaer (1905, 1934) and the concept of potential evapo-transpiration as a basis for climate classification suggested by Thornthwaite (1931, 1948).

The Danish botanist, Christen Raunkiaer, proposed a classification system for vegetation based on the location of buds that protected the delicate meristems of plants during the inclement time of the year. These buds are referred to as renewal buds and their survival is key to the survival of the plant (Fig. 3.4). Raunkiaer pointed out that the preponderance of renewal buds in arctic areas was produced at the surface of the ground where they were insolated from low winter temperatures by snow. Renewal buds above the level of the snow would be subjected to freezing, as would renewal buds beneath ground in areas where the permafrost reached the

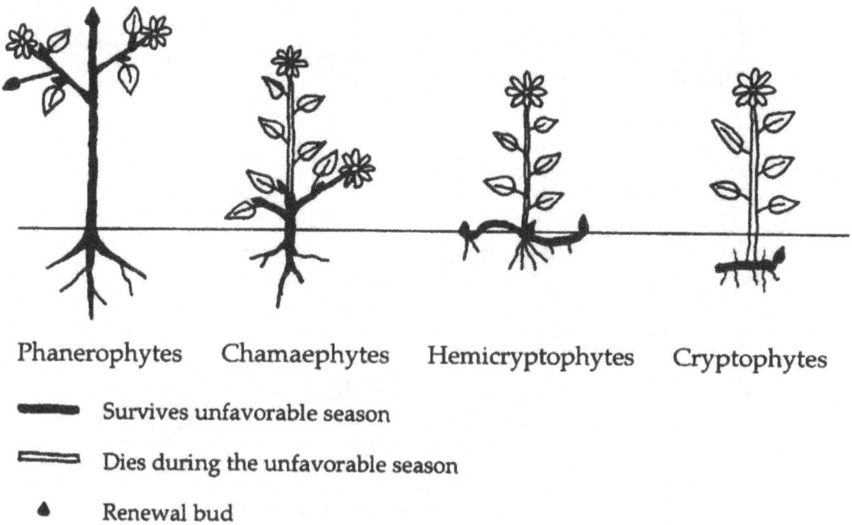

Fig. 3.4 Raunkiaer's life-form classification (Redrawn from Raunkiaer, C. 1934)

soil surface. He referred to plants with their renewal bud at the surface of the ground as hemicryptophytes. In contrast to the hemicryptophytes were the phanerophytes whose renewal buds were at the ends of the aerial branches well above the ground surface. In general, phanerophytes do not occur in arctic regions.

Although Raunkiaer's classification did not lead to new directions in the mapping of the world's vegetation, its did contribute to our knowledge of the adaptations in plants morphology to different climatic regions. It has also been the basis for understanding the success and/or failure of the introduction of plant into different urban environments. As a concept, Raunkiaer's life form classification underlies the later development of frost hardiness zone maps used by arborists in the selection of perennial plants for urban environments (Fig. 3.5).

Charles Thornthwaite, an American climatologist, proposed a new climate classification in 1931 based on four factors: moistness, temperature, seasonality of moisture, and temperature regime (Thornthwaite 1931, 1948). His novel concept of moistness was measured by a moisture index computed from the annual water surplus and the annual water deficit based on potential evapo-transpiration. This resulted in a climate classification system that was more sensitive to the growth of plants than the classification proposed by Köppen. Thornthwaite showed that local climates could be depicted by climatic diagrams to illustrate periods of water surplus and water deficit for plant growth based on the annual pattern of precipitation and evapo-transpiration (Thornthwaite and Hare 1955). The utility of these diagrams, with reference to plant growth, was in their ability to show that portion of the year in which a water deficit occurred.

The contrast in water surplus and deficit is seen in climate diagrams for New York City, Cairo, and Athens (Fig. 3.6). These diagrams have been modified from the

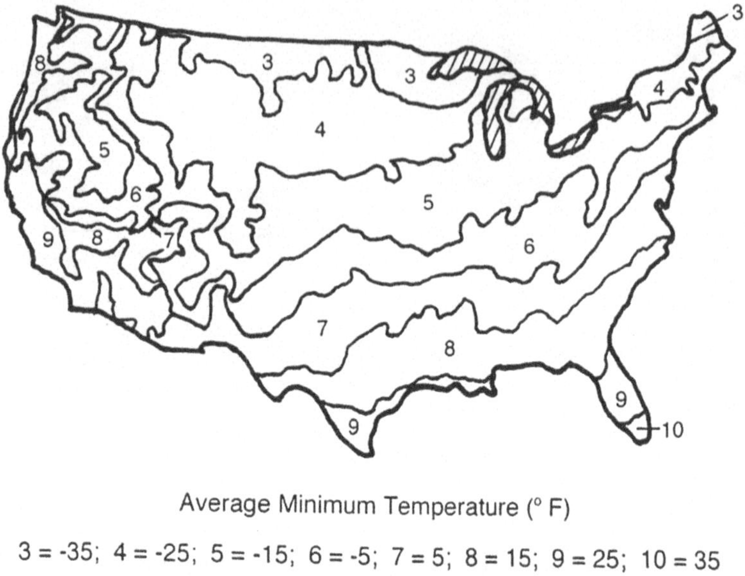

Fig. 3.5 Frost hardiness zone map of the conterminous United States

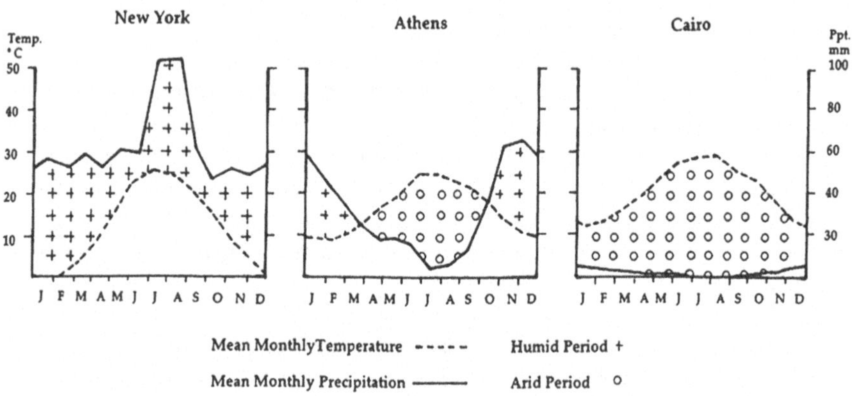

Fig. 3.6 Climate diagrams for New York City, Athens, and Cairo

climate diagrams published by Walter et al. (1975) to emphasize the periods of water surplus (humid period) and water deficit (arid period). In the diagram for New York one sees a water surplus throughout the year. New York receives at least 48mm (1.9 inch) of precipitation every month that maintains a water surplus. The soil does not become parched at any time during the year and the water surplus is at a maximum during the growing season (June, July, and August). The pattern of water surplus through the year is typical of other locations in the deciduous forest biome. The climate diagram for Cairo indicates a water deficit throughout the year.

In Cairo average monthly precipitation rarely reaches 10mm (0.4 inch). The water deficit is at a maximum during the warmest months of the year (June, July, and August) when one might expect the greatest plant growth due to higher levels of solar radiation. However, the water deficit restricts plant growth at this time. Other locations in the desert biome share a yearlong pattern of water deficit with Cairo. The climate diagram for Athens shows periods of both water surplus (October to April) and a deficit (May to September). High rainfall in the autumn and winter months provide moisture for a water surplus. As rainfall decreases in spring and summer, the rising temperatures lead to a water deficit that is characteristic of both Athens and other locations in the Mediterranean woodland/scrub biome. Thornthwaite's climatic classification, based on potential evapo-transpiration, has proven useful in understanding the limits on tree growth in certain biomes and the seasonal need for irrigation of trees used in these biomes.

By the middle of the twentieth century the widespread use of aerial photography allowed for the drawing of more accurate maps of the world's vegetation. These vegetation maps have been used to delineate the world's biomes. Five classifications systems for the world's biomes (Finch and Trewartha 1949; Bartholmew 1950; Hammonds 1961; Eyre 1968; Breckle 2002) were reviewed for the purpose of selecting an appropriate classification system for this book. Table 3.1 identifies the biome designations used in each of the five classifications.

In developing a classification system appropriate for this book a decision was made to produce a classification that would use each of the major physiognomic types as major categories and to minimize the number of subcategories of each types. To that end the following classification system was adopted: tropical forest, broadleaf evergreen forest, deciduous forest, coniferous forest, mixed deciduous-coniferous forest, Mediterranean woodland/scrub, savanna/scrub, grassland, desert, tundra, and highland (Table 3.2).

The classification system adopted here merges the tropical rain forest identified by each of the five sources (Finch and Trewartha 1949; Bartholmew 1950; Hammonds 1961; Eyre 1968; and Breckle 2002) with the tropical seasonal forest (Eyre 1968) and the tropical deciduous forest (Breckle 2002). This merger was justified on the absence of the later categories from the first three sources (Finch and Trewartha 1949; Bartholmew 1950; Hammonds 1961) and the significance of high annual rainfall and minimal variation in monthly temperature in both categories.

The term "broadleaf evergreen forest" will be used in place of the terms subtropical forest (Hammonds 1961) and temperate evergreen forest (Breckle 2002). Broadleaf evergreen forest more directly conveys the appearance of this biome and avoids confusion with coniferous forests in the temperate zone. Although the broadleaf evergreen forest was not identified by three of the sources (Finch and Trewartha 1949; Bartholmew 1950; Eyre 1968), it is a recognizable biome and is important to include to distinguish the vegetation of a considerable area of Japan, the coast of the Asian mainland, and portions of Australia.

Both the coniferous forest and mixed deciduous-coniferous forest types are recognized by all five sources and are included in the classification of biomes used here. The deciduous forest type was proposed by Eyre (1968) and Breckle (2002).

Table 3.1 Comparison of biome classification used by Finch and Trewartha (1949), Bartholmew (1950), Hammonds (1961), Eyre (1968), and Breckle (2002)

Finch and Trewartha (1949)	Bartholmew (1950)	Hammonds (1961)	Eyre (1968)	Breckle (2002)
Tropical Rainforest	Tropical Rainforest	Tropical Rainforest	Tropical Rainforest	Evergreen Tropical Rainforest
Semi deciduous Tropical Forest			Tropical Seasonal Forest	Tropical Deciduous Forest
Scrub and Thorn Forest	Woodland and Brush	Scrub and Thorn Forest	Semi-Desert Scrub and Woodlands	
Mediterranean Scrub Forest	Mediterranean Dry Type	Mediterranean Mixed Forest	Sclerophyllous Formations	Sclerophyllous Woody Plants
Broadleaf and Mixed Broadleaf–Coniferous Forest	Mixed Deciduous Forest	Temperate Forest	Mixed Forests of Middle Latitude	Broadleaf Deciduous Forest Boreal Coniferous Forest
		Subtropical Forest		Temperate Evergreen Forest
Coniferous Forest	Coniferous Forest	Coniferous Forest	Coniferous Forests	Boreal Coniferous Forest
			Broad-Leaved Forests of Middle Latitude	Broadleaf Deciduous Forest
Savanna	Grassland	Savanna	Tropical Savanna	Tropical Savanna
Prairie		Prairie	Prairie	
Steppe		Steppe	Steppe	Steppe
			Extra Tropical Grasslands of the Southern Hemisphere	
Desert Scrub	Arid and Semi-Desert	Desert Scrub and Waste	Microphyllous Forest, Semi-Desert and Desert	Subtropical Desert
Tundra	Tundra	Tundra and Alpine	Tundra and Alpine	Tundra
Undifferentiated Highland	Mountain Type	Unclassified Highlands		Mountains
		River Valleys and Oasis		

Table 3.2 Classification of biomes

Physiognomic type	Biome
Forest	Tropical Forest
	Broadleaf Evergreen Forest
	Deciduous Forest
	Coniferous Forest
	Mixed Deciduous-Coniferous Forest
Woodland	Mediterranean Woodland/Scrub
Savanna	Savanna/Scrub
Grassland	Grassland
Scrub	Desert
Tundra	Tundra
Mixed	Highland

It is subsumed in the mixed deciduous-coniferous forest type (and variations on this term) by the other sources. The decision to distinguish this type from the broader mixed deciduous-coniferous forest was based on the distinct appearance of forests composed of deciduous trees and the correlation of this forest type to slightly milder temperatures and higher rainfall than the mixed deciduous-coniferous forest.

The Mediterranean woodland/scrub was identified in all five sources. It is the biome surrounding the Mediterranean ocean and is characterized by a mosaic of evergreen woodlands and scrublands of sclerophyllus shrubs well adapted to the wet winter/dry summer climate of the Mediterranean region. Fragments of the Mediterranean woodland/scrub also occur in South Africa, California, Chile, and Australia.

The term "savanna/ scrub" is used in the current classification system to include both the savanna, scrub, and thorn forest types recognized by the five sources. These were combined on the basis of the gradient that occurs between these types in South America, Africa, and India. In many areas the ecotones separating the savanna, scrub, and thorn forest are often more extensive than the pure types themselves. Also, the maps presented in the five sources (Finch and Trewartha 1949; Bartholmew 1950; Hammonds 1961; Eyre 1968; and Breckle 2002) often did not agree on the distribution of the types.

Grassland was adopted as a term to include all grass dominated vegetation types (grassland, prairie, steppe, tropical grassland) used by the five authors. Although there is variation in the climate across these different types of grass dominated landscapes, these landscapes are characterized by a very limited tree flora.

The desert and tundra types are identified in all five sources. These two types are clearly defined on the basis of climate. The desert biome can be defined as an area of high temperature that is in water deficit throughout the year. The tundra is defined by its low average annual temperature, extremely low winter temperature, and low rainfall. Both biomes have distinctive vegetation as a result of these climatic limits.

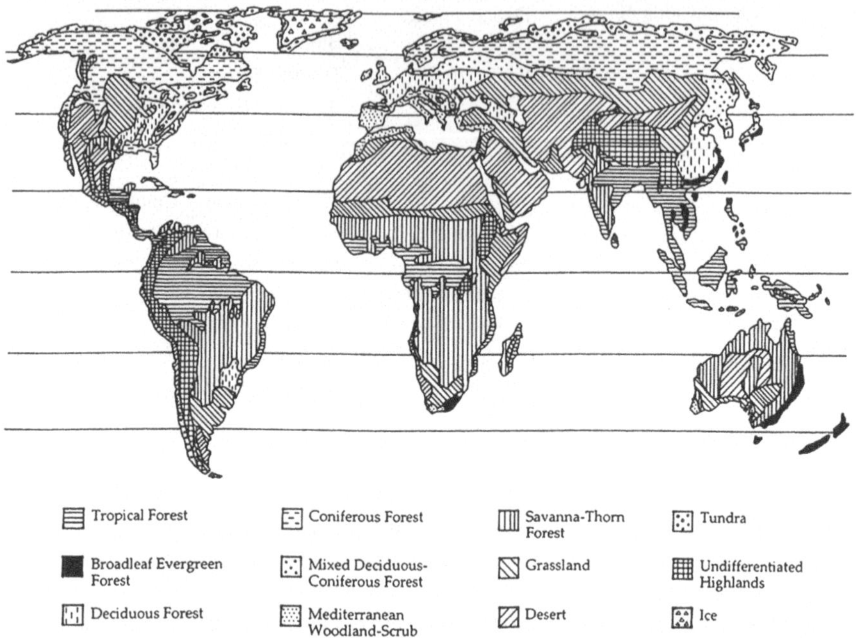

Fig. 3.7 Distribution of biomes

The final classification type is highland. This unit of classification was recognized by four of the five sources, while two of these (Bartholmew 1950; Breckle 2002) used the term mountains or mountain type for this category. It represents of a challenge for biome classification because it is a large geographical region in which the climate and vegetation are controlled by elevation and the lower elevation limit of the unit varies with latitude. For example, "highland" on the Kenai Peninsula of Alaska (latitude 60° N) begin at an elevation of 1500 to 2000 feet. In contrast, "highland" is at or above 10,000 feet in the mountains of Ecuador near the equator. At higher elevations there tends to be a small-scale mosaic of vegetation types that may include sub-alpine forests, scrub, grasslands, and tundra. One cannot easily parcel these local mountain vegetation types into the larger biomes because of their geographical discontinuity with the biomes occurring at lower elevations. The influence of elevation is a dominant factor determining both the climates and vegetation of these highland areas. Elevation trumps the influence of latitude and global wind patterns that are dominant in determining the distribution of climate and vegetation at lower elevations.

The distribution of the biomes is shown in Fig. 3.7. The biomes have a tendency to follow the latitudinal patterns of temperature belts in the northern hemisphere suggested by von Humboldt (1805) and Köppen (1931). These temperature belts are somewhat congruent with the sequence of the tundra, coniferous forest, mixed deciduous-coniferous forest, deciduous forest, grassland, and desert biomes as one move south from the cold polar regions to warmer belts approaching the equator. This pattern is interrupted by variations in season patterns of rainfall that account for the Mediterranean woodland/scrub, savanna/ scrub, and broadleaf evergreen biomes. The high rainfall in equatorial regions of South American, Africa, and Southeast Asia accounts for the pattern of distribution of the tropical forest biome.

Human Settlement in the World's Biomes

Most authorities agree that permanent human settlements evolved in response to the development of technologies for the production and storage of surplus food (Adams 1960; Mumford 1961; Sjoberg 1965; Kostof 1991; Diamond 1997). The capacity to accumulate and store food allowed people to live in permanent communities. The ancient Chinese village of Pan-po-Ts'un (Fig. 3.8), near the present city of Xian, built about 5000 BC consisted of 41 pit houses and 300 below ground storage pits (Chang 1986). These early communities were initially supported by agriculture and fishing, but trading of commodities soon allowed people to live in cities along trade routes once technologies for food storage had been developed.

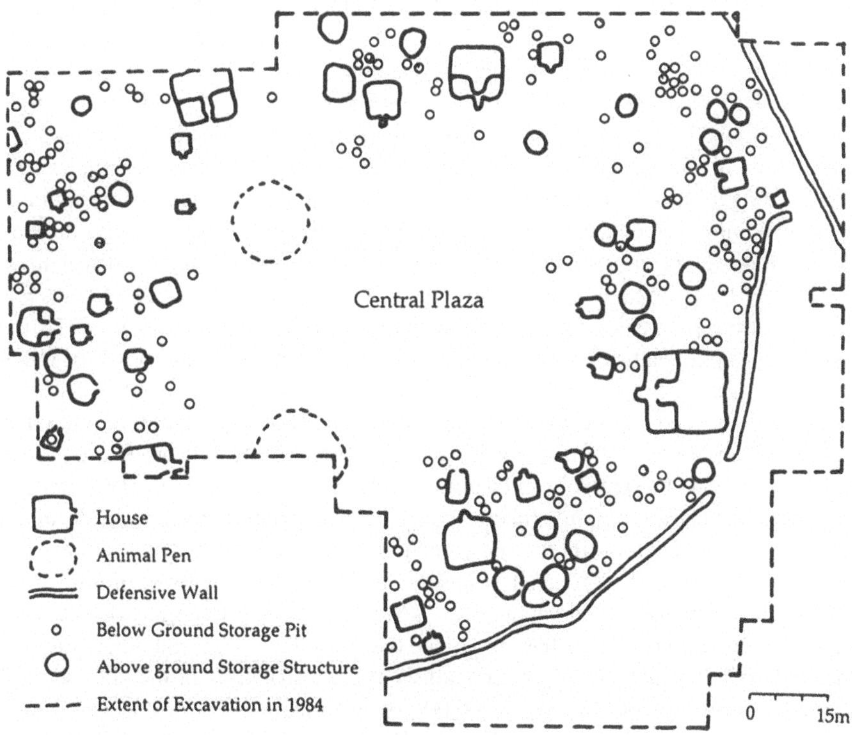

Fig. 3.8 Ancient Chinese village of Pan-po-Ts'un

Various explanations have been advanced to explain the further evolution of cities once technologies for the production and storage of surplus food been realized. These theories have for the most part focused on the functional role of the city. Aurousseau (1928) organized the dominant functions of cities into the six following classes: (1) administration, (2) defense, (3) culture, (4) production (manufacturing), (5) communication (commerce), and (6) recreation. Transportation often played an important role in the location of cities as can be seen by the location of ancient cities along rivers or on overland trade routes.

The development of the city cannot be separated from the evolution of civilization. In the early part of the twentieth century James Fairgrieve (1921) suggested that regional variations in civilizations were based on human control over energy. In Fairgrieve's view the distribution of civilization can be explained in terms of the control and utilization of energy. In the early part of the twentieth century coal was the principle source of energy for industrial production and urban consumption in Europe and North America. If one uses the existence of large cities as a measure of civilization, one would see a general correlation between the locations of coalfields and major cities in Europe, North America, and China in the early nineteenth century (Finch and Trewartha 1936).

Ellsworth Huntington (1924) proposed an alternative theory to Fairgrieve's 'control of energy' theory that was based on climate. Huntington suggested that the distribution of civilization was correlated with climate. He concluded that temperature, humidity, and the variability in these factors are of greatest importance affecting human health and energy. Huntington's view was based in part on interpretation of the lack of advancement of civilization in areas of "unsolved climatic problems". These included (1) the cold regions of the sub arctic and tundra, (2) the wet tropics, and (3) the dry lands. Huntington's theory was rejected by many geographers as representing a Euro-American centric view of civilization. If one limits the application of Huntington's concept to Europe and North America one finds a high correlation between the occurrence in 1920 of larger cities (> 500,000) and the distribution of Huntington's "high climatic energy" zone. It may be argued that large cities are not a measure of civilization, but the evolution of urban life has often been suggested as a significant step forward in the rise of civilizations (Mumford 1961; Hall 1998).

Toward the end of the twentieth century Jared Diamond (2004) proposed a hypothesis in his book "Collapse: How societies choose to fail or succeed" that touches on both Fairgreive's and Huntington's concepts. Since the 1920s when Fairgrieve's and Huntington's theories were being argued, technological developments have resolved the "unresolved climatic problems" and new sources of energy have been tapped and more easily transported. The meteoric growth of cities like Phoenix, Arizona has depended upon the advent of air conditioning while development of petroleum resources has stimulated the growth of Dubai and other cities on the Arabian Peninsula. However, an analysis of the distribution of the 404 cities of the world with populations over 1,000,000 in 2000 suggests that whatever resolution of the "unresolved climatic problems" that has occurred since the 1920s has not resulted in a uniform distribution of large cities across the world's biomes. The

Table 3.3 Distribution of the world's cities with populations over 1,000,000 in the various biomes

Biome	Area (million mi.^{2}	Percent of world's land mass	Number of cities > 1,000,000	Percent of number of cities >1,000,000	Density of cities (#/1 million sq. mi.^{2})
Tropical Forest	7.5	13	93	23	12
Broadleaf Evergreen	0.8	1	23	6	29
Deciduous Forest	3.0	5	75	18	25
Coniferous Forest	8.6	15	9	2	1
Mixed Deciduous-Coniferous Forest	2.0	4	48	12	24
Mediterranean Woodland/Scrub	1.0	1.2	35	9	35
Savanna/ Thorn Scrub	10.6	19	40	10	4
Grassland	7.4	13	49	12	7
Desert	9.6	17	23	6	2
Tundra	3.7	6	0	0	0
Highland	2.4	6	9	2	4

coniferous forest, savanna/ scrub, desert, tundra, and highland biomes each have a relatively low density of cities with populations over 1,000,000 (Table 3.3). The higher densities of large cities in the broadleaf evergreen forest, deciduous forest, mixed deciduous-coniferous forest, and Mediterranean woodland/scrub may have more to do with the historical development of the resources of biomes, international trade, and the growth of human populations than with the effects of Huntington's "climatic energy" on the inhabitants. It is interesting to speculate that wood as a fuel and a raw material was important to the early growth of human populations and cities in certain biomes. If wood alone was the critical factor one would expect to see a higher density of large cities in the coniferous forest biome. That this is not the case may be due to the limitations imposed by the climate in the northern portions of the coniferous forest biome on agriculture.

Thirty-three cities were selected for a detailed study of their urban forests. The distribution of these cities in the various biomes is shown in Table 3.4 and in Fig. 3.9. An attempt was made to select cities with populations over 1,000,000 in each biome. It was assume that cities of this size would have an economic base sufficient to provide funds for urban forest planning and management. It was necessary to select some cities with populations of less than 1,000,000 because of the absence of large cities in certain biomes, such as the tundra biome.

Variations in the historical and cultural development of cities are also addressed in the selection. For example, the three cities selected in the tropical forest biome (Bangkok, Singapore, Rio de Janeiro) have quite different cultural and historical backgrounds. Bangkok was established as the capital of Thailand in 1872 by King Ramathibodi. At that time a small village surrounded two ancient temple complexes on a bit of high ground occurred along the Chao Praya River. This site was selected for the future capitol city because of the strategic advantages of the higher ground.

Table 3.4 Cities selected for study and their population

Biome	City	Population[a]
Tropical Forest	Bangkok	5,584,200
	Rio de Janeiro	5,533,011
	Singapore	4,300,419
Broadleaf Evergreen Forest	Hong Kong	6,977,000
	Sydney	3,713,000
	Tokyo	8,130,000
Deciduous Forest	London	7,007, 100
	New York	7,420,166
	Paris	2,125,000
Coniferous Forest	Jacksonville	735,617
	Seattle	563,374
	Vancouver	2,117,000
Mixed Deciduous-Coniferous Forest	Beijing	6,970,000
	Moscow	8,434,000
	Stockholm	703,627
Mediterranean Woodland/Scrub	Athens	3,000,000
	Los Angeles	3,694,820
	Tel Aviv	1,236,000
Savanna/Thorn Forest	Brasilia	1,800,000
	Lagos	7,937,000
	New Delhi	7,174,755
Grassland	Buenos Aires	12,960,976
	Johannesburg	5,700,000
	Tehran	6,750,000
Desert	Cairo	6,955,000
	Dubai	647,100
	Phoenix	1,321,045
Tundra	Murmansk	600,000
	Nuuk	15,000
	Reykjavik	201,000
Highland	Cusco	359,000
	Lhasa	373,000
	Quito	1,500,000

[a]Population for the city, metropolitan areas surrounding each city have additional populations. For example the population of the Tokyo metropolitan area was 26.4 million in the year 2000 (Wright 2003)

King Ramathibodi was able to keep various European countries that were interested in colonizing Thailand from realizing their ambitions. As a result, Bangkok developed without direct European control of city planning and urban design. Singapore, in contrast, was a city developed under the influence of British colonialism during the same period that Bangkok came into being. Its pattern of streets and public parks echo patterns of urban development used in England. Much earlier, Rio de Janeiro

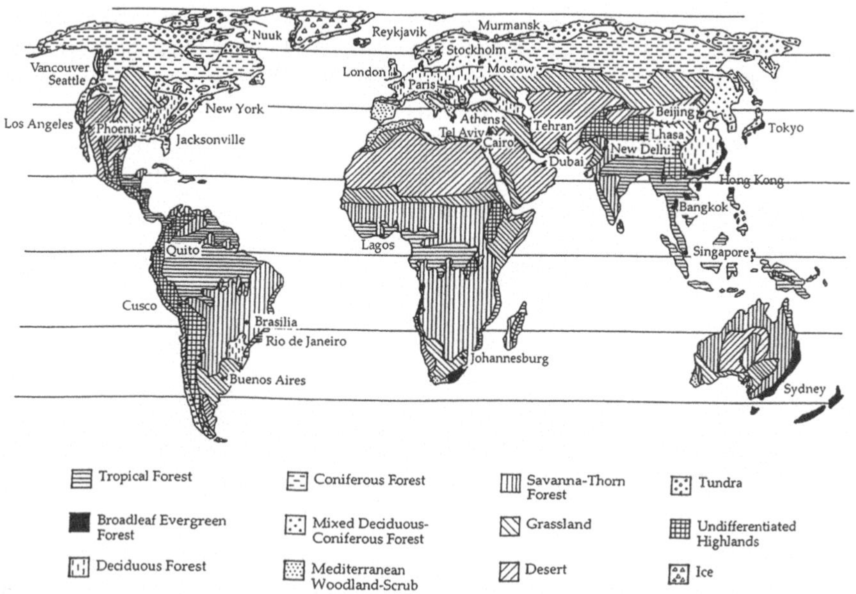

Fig. 3.9 Location of cities in biomes

Table 3.5 Dominant Tree Species in the Deciduous Forest Biomes of North America and Europe

North America[a]	Europe[b]
American beach – *Fagus grandifolia*	Ash – *Fraxinus excelsior*
Basswood – *Tilia americana*	Beech – *Fagus sylvatica*
Black oak – *Quercus velutina*	Field maple – *Acer campestre*
Green ash – *Fraxinus americana*	Hornbeam – *Carpinus betulus*
Red Oak – *Quercus rubra*	Pedunculate oak – *Quercus robur*
Shagbark hickory – *Carya ovata*	Sessile oak – *Quercus petraea*
Sugar maple – *Acer saccharum*	Silver birch – *Betula pendula*
White oak – *Quercus alba*	Small leaf lime – *Tilia cordata*
Yellow poplar –*Lirodendron tulipifera*	Small leaved elm – *Ulmus minor*

[a]Delcourt and Delcourt (2000)
[b]Polunin and Walters (1985)

was established as a port city by the Portuguese. It became the capitol of Portugal from 1808 to 1821 when the Portuguese royal family fled Europe to avoid the conquest of the Iberian Peninsula by Napoleon. Considerable investment was made in the construction of buildings and avenues during the King's residence in the city. However, major development followed independence from Portugal in 1822 and much of the colonial character of the city was erased.

A final criterion in the selection of cities was to select cites from each biome on different continents. This criterion enables one to investigate the importance of local tree species in the development of urban forests on different continents within the same biome. For example, the question of the use of local species was addressed in a comparison of New York City, London and Paris. All three of these cities occur in the deciduous forest biome, but the tree species composition of the biome in North America is different than that of the British Isles and continental Europe (Table 3.5). It should be noted that each of the species listed for the deciduous forest biome in North American were observed in the urban forest of New York City as were each of the European species observed in Paris.

Chapter 4
History

Introduction

Human history is an ever-widening trace left by the modification of the earth's landscape and its vegetation. This trace is particularly evident in the evolution of cities. The builders of early cities in forested regions were compelled to remove trees, while trees were introduced into cities built in grasslands and other treeless areas. The treatment of natural vegetation at the sites of human settlement and the introduction of exotic vegetation, especially trees, is an often-overlooked part of human history. Early human settlements were characterized by the general elimination of trees within the protective walls. During periods of greater safety people modified their settlements by the planting of trees. The use of trees in human settlements was closely linked to various social, economic, and technological developments influencing the evolution of cities. Knowledge of history of trees in urban areas can provide a better understanding and deeper appreciation of the urban forest. It also provides a context for future design and management of trees in urban areas.

Early Human Settlements

The capacity of people to modify local landscapes was initially very limited by technology and population size. The mastery of fire in Paleolithic times provided a technology for changing vegetation. Fire could be used to burn away unwanted plants at local sites were settlements were to be established. Fire is still used today by indigenous people in northern Australia to eliminate grass and herbs from locations chosen for campsites (Nicholson 1981). Burning off of the vegetation by these people is done to drive away poisonous snakes and eliminate insects. Native Americans living in parts of southern California used fire to maintain brush-free zones around their villages as a means of fire protection (Lewis 1973).

J.R. McBride, *The World's Urban Forests*, Future City 8,
DOI 10.1007/978-3-319-52108-4_4

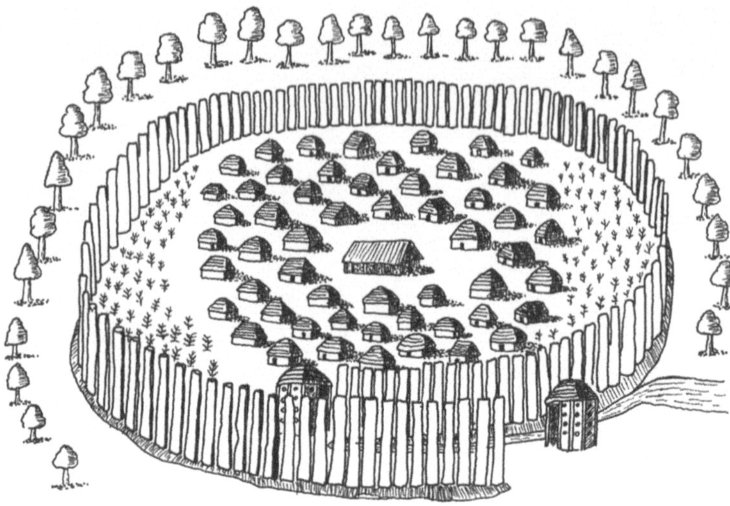

Fig. 4.1 Timucuan (Redrawn from Jacques le Moyne, 1564)

In the Neolithic age the development of agriculture, fishing, lithic industries, and economies based on trading made it possible to establish and sustain permanent communal living sites. Communal living, in groups of people larger than the extended families of the Paleolithic period, intensified the modification of the landscape. The improved tool making skills of the Neolithic age provided people with more efficient tools for the felling of trees and building structures. Some trees at early village sites were no doubt removed to provide space and construction materials for structures.

The ancient farming village of Pan-p'o-Ts'un, near the present city of Xi'an in central China, supplies evidence of the capacity of Neolithic people to modify local forests. Radiocarbon tests suggest that the earliest houses date from about 5000 BC. It can be assumed that trees on the site were removed during the building of Pan-p'o-Ts'un since this village was established in an area of hardwood forest. Pollen analysis of Pan-p'o-chan deposits show that pollen of forest trees decreased dramatically with the occupancy of the site. This decrease is attributed by Hay (1973) to the cutting of trees with polished stone adzes commonly found at the site. Archaeological excavation of Pan-p'o-Ts'un revealed 41 pit houses, 300 below ground storage pits, and no evidence of trees occurring within the confines of the village.

Paintings and drawings by early European and American explorers of the North America indicate a general absence of trees and other plants within the confines of indigenous villages. Jacques Le Moyne, a French cartographer on the Laudonniere expedition to Florida in 1564, painted a Timucuan town in what is now Florida (Lorant 1946). His painting shows a village surrounded by a palisade of logs (Fig. 4.1). Within the perimeter of this palisade no trees occur. Outside of the

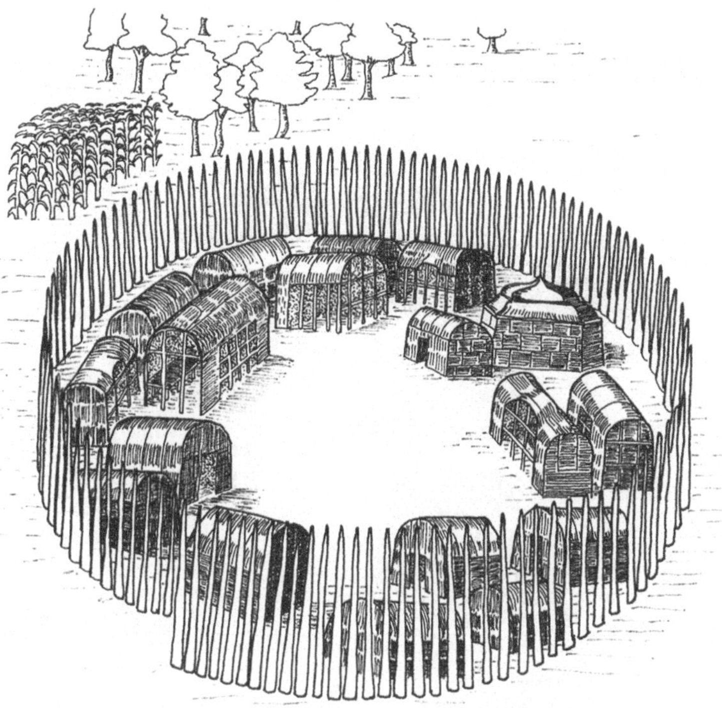

Fig. 4.2 Pomeiooc (Redrawn from John White, c 1580)

palisade one sees the edge of the surrounding forest. What might be interpreted as a crop plant occurs within the palisade.

In contrast, a painting by John White done in the late 1580s of Pomeiooc, a village in what is now North Carolina, shows no plants within the palisade (Fig. 4.2). In similar a fashion George Catlin's 1823 painting of a Mandan village on the Missouri River shows no vegetation within the confines of its palisade (Catlin 1837–39). Palisades and walls of earth or stone were built around villages as defensive measures. Space for trees and vegetable gardens within these defensive perimeters was at a premium. A similar absence of trees and garden areas is often seen in paintings and drawings of castles and fortified cities in Europe in the early Middle Ages (Crisp 1924).

Secoton, the principle town of the Secotan people of the North Carolina, was also painted by John White during his visit to the new world (Fig. 4.3). Unlike his painting of Pomeiooc, Secotan is not surrounded by a wall of logs. Secoton had trees and areas devoted to the raising of corn adjacent to the dwellings. The absence of a palisade around Secotan suggests a more secure location that was free from attacks by neighboring people. Trees could be incorporated into the village to provide shade and fruit, while crops such as the corn illustrated in White's painting could be easily maintained adjacent to individual houses.

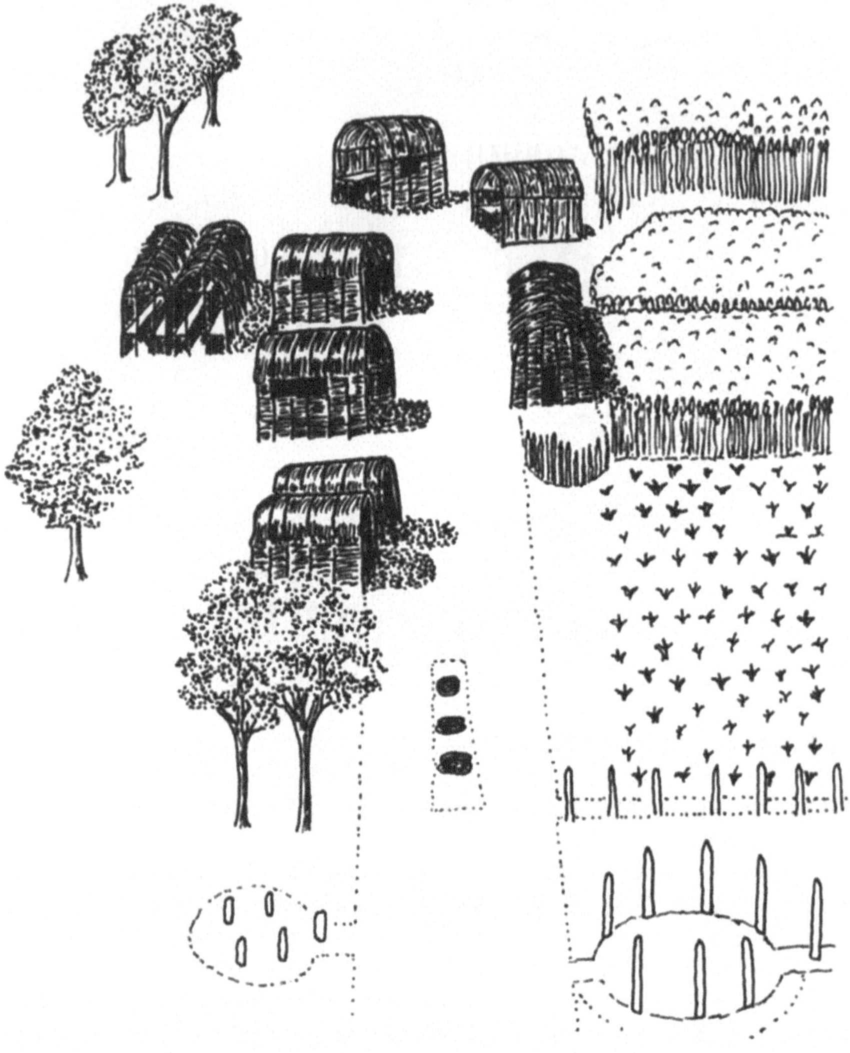

Fig. 4.3 Secoton (Redrawn from John White, c. 1585)

The image of Secoton is not unlike present day indigenous villages in the Amazon Basin of South America. Quechuan villages in the Amazon region of Ecuador included fruit-bearing trees and small kitchen gardens. Trees present on the sites prior to construction of the villages, or planted subsequently, are maintained as a source of fruit. Small kitchen gardens are planted adjacent to the houses, while larger areas for crop production were cultivated at the margins of the village.

Treatment of vegetation at village sites, as illustrated in the drawings and paintings of European and American explorers of North America, and as can be observed today in the villages of indigenous people living in the Amazon, demonstrates a common practice of modifying local sites for human habitation. Modifications commonly involve the removal of trees and other types of natural vegetation to provide space for buildings, supply building materials, and for hazard reduction. Vegetation modification in these early settlements may also involve the maintenance or establishment of some trees for fruit crops and the introduction of small kitchen gardens. The need to modify local vegetation at sites of human habitation continued to be addressed as social interactions, economics, and technology lead to larger and larger urban settlements.

Large Urban Areas in Mexico at the Time of the Spanish Conquest

Large urban areas occurred in Mexico, Central, and South American at the time of the first European exploration of the western hemisphere. The characteristics of trees and other types of vegetation in these early metropolitan areas of the new world may serve as a key to understanding the role trees and other plants played in the ancient metropolitan areas of Africa, the Middle East, and China. The largest city in the western hemisphere at the time of the early European exploration was the Aztec metropolis of Tenochtitlan (now Mexico City). With a population of 250,000, it was larger than any city in Europe in the sixteenth century (Josephy 1994). A map of Tenochtitlan, published in Europe in 1524 shows seven large gardens within the city (Fig. 4.4). These gardens occupied 6.5% of the area of Tenochtitlan. As points of reference, urban green space in Manhattan, the island borough of New York City, amounts to 10.2% today.

Prescott (1937) pieced together the following views of the gardens of Tenochtitlan and the nearby city of Iztapalapan based on the observation of Cortez, Diaz, and Martyr:

> Tenochtitlan
> The principle street (of Tenochtitlan), conducting from the causeway, penetrated in a straight line the whole length of the city, and afforded a noble vista, in which the long lines of low stone edifices were broken occasionally by intervening gardens, rising on terraces and displaying all the pomp of Aztec horticulture...Extensive gardens were spread out around these buildings, filled with fragrant shrubs and flowers, and especially medicinal plants...Amidst this labyrinth of sweet-scented groves and shrubberies, fountains of pure water might be seen throwing up their sparkling jets and scattering refreshing dew over the blossoms.

> Iztapalapan
> But the pride of Iztapalapan, on which its lord had freely lavished his care and his revenues, was its celebrated gardens. They covered an immense tract of land; were laid out in regular squares, and the paths intersecting them were bordered with trellises, supporting creepers and aromatic shrubs that loaded the air with their perfumes. The gardens were stocked with fruit-trees, imported from distant places, and with the gaudy family of flowers which belonged to the Mexican flora, scientifically arranged, and growing luxuriant in the equable temperatures of the table-land. The natural dryness of the atmosphere was counteracted by means of aqueducts and canals that carried water into all parts of the ground.

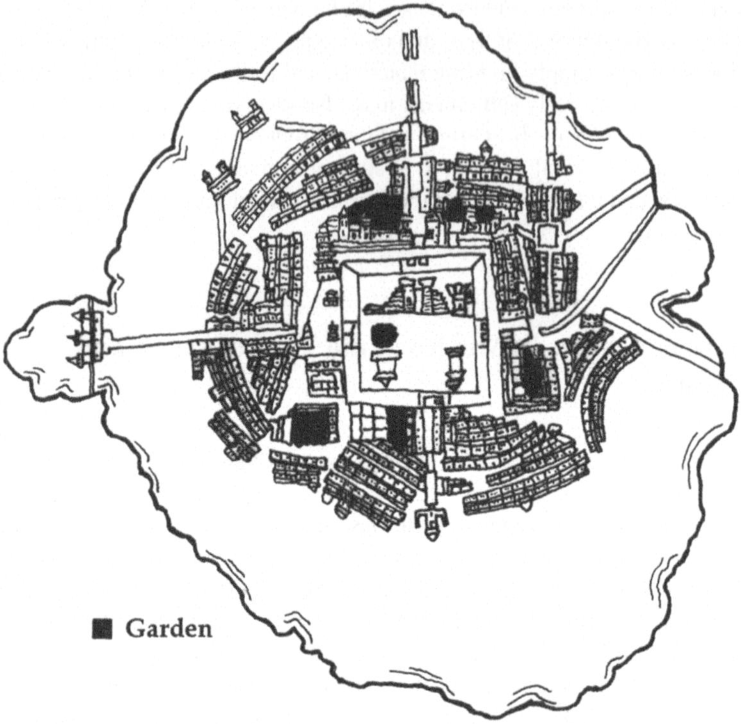

Fig. 4.4 Tenochtitlan (Redrawn from a map published in 1524, Josephy 1994)

The gardens in Tenochtitlan, Iztapalapan, and other large urban areas of the western hemisphere were established and maintained for the pleasure of the ruling class and for the medicinal value of certain plants. Gardens for the pleasure of kings and their wealthy contemporaries were also built in the ancient cities of Africa, the Middle East, and China. Gardens for the study of plants, like those found in Iztapalapan, also occurred in some of these ancient cities. The earliest descriptions of Tenochtitlan and Iztapalapan indicate that the common man did not have space for a garden adjacent to the closely packed houses in these cities. The protection of these large urban populations during the almost continuous warfare during the Aztec period dictated against the inclusion of open space areas within the defensive perimeters of cities.

Urban Trees in Ancient Egypt

The earliest description of urban vegetation in the eastern hemisphere is from the hieroglyphics in the tomb of Methen, governor of the northern delta district of Egypt in the twenty-sixth century B.C. (Cowell 1978). These hieroglyphics rather tersely describe the garden of Methen as follows:

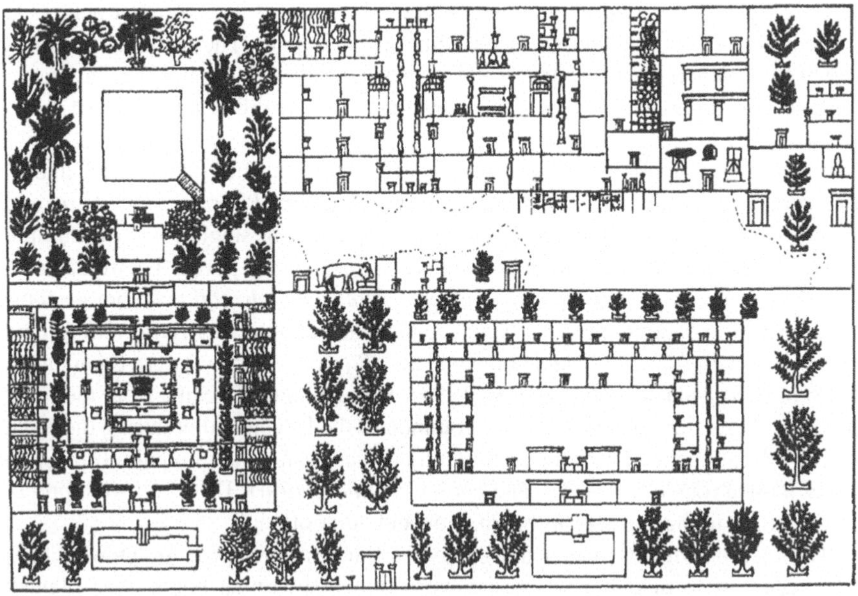

Fig. 4.5 Akhenaton's palace and garden (ca. 1370 BC) (Redrawn from Dixon 1969)

> Two hundred cubits[1] long and 200 cubits wide; built and equipped; fine trees were set out; a very large lake was made therein; figs and vines were set out. Very plentiful trees and vines were set out; a great quantity of wine was made therein.

In the climate of Egypt shade was sought to provide relief from the desert sun. The consumption of fruit was also valued for its moisture content when fresh and probably served as an important means of maintaining electrolytes in a climate that promoted perspiration. In recognizing the functional value of these trees one should not overlook their ornamental value. The great sense of symmetry exhibited in the design of many of the temples of ancient Egypt is also present in the design of the gardens maintained by the Pharaohs and their high-ranking administrators.

Representations of gardens were carved onto the walls of the tombs of Egyptian rulers. The plan for the palace and garden Akhenaton (1379–1362 BC) shows a garden dominated by trees surrounding a square pool. The representation of the base of the trees suggest they are growing out of the soil on which the garden had been planted (Fig. 4.5). The plan for Akhenaton's palace and garden also shows a considerable number of trees that are growing in pits individually surrounded by low walls around the palace and the houses of his officials. Archaeological excavations at several Egyptian sites indicated that tree pits were dug into the bedrock adjacent to houses and other structures to accommodate the roots of trees. These pits were supplied water by surface irrigation ditches. Since the 1970s a number of

[1]Ancient measurement based on the length of the forearm from the elbow to the tip of the middle finger and usually equal to 1.5 feet.

archaeologists have focused their attention on the spaces surrounding structures at ancient ruins in the Middle East and Mediterranean regions. The work of Jashemski (1979) in Pompeii, Kemp (1989) in Egypt, and Carroll-Spillecke (1992) in Greece has added to our understanding of the inclusion of trees in urban areas of the Ancient world.

The representations of gardens on the walls of tombs and papyrus documents depict the use sycomore fig (*Ficus sycomorus*), date palm (*Phoenix dactylifera*), and doum palm (*Hyphaene thebaica*). These were planted in single rows or multiple rows to form a grid of trees around square or rectangular pools. The remains of fruits and nuts of six tree species (persea – *Minusops laurifolia*, sycomore fig – *Ficus sycomorus*, doum palm – *Hyphaene thebaica*, almonds – *Prunus dulcis*, date palm – *Phoenix dactylifera*, and pomegranate – *Punica granatum*) were found in the tomb of King Tutankhamen (Hepper 1990). These assumedly came from the royal gardens. Other trees whose fruit was common to the Egyptian diet during the reign of King Tutankhamen were the common fig (*Ficus carica*), argum palm (*Medemia argun*), olive (*Olive europaea*), and cordia (*Cordia gharaf*). These too were probably common to the gardens of the Pharaoh and other officials.

Even minor officials in ancient Egypt were able to afford space for a few orchard trees adjacent to their houses. A reconstruction of Egyptian houses and planted courtyards in Amarna (c. 1350–1334 BC.) suggests that various Egyptian officials maintained fruit trees within walled enclosures adjacent to their houses (Carroll 2003). Paintings on the tomb walls on Egyptian officials frequently show sycomore fig, common fig, date palm, and pomegranate. In addition to these fruit trees, willow (*Salix subserrata) and* tamarisk (*Tamarix apylla)* are recorded in the tomb of Ineni (sixteenth and fifteenth century BC) at Thebes as being present in his garden (Dziobek 1992). A delightful wooden model from the tomb of Meketre (an official of the eleventh Dynasty, serving in the reign of Mentuhotpe II [2061–2010 B.C.]) shows a rectangular pool surrounded on three sides by sycomore trees (Fig. 4.6). Echoes of this tradition of incorporating fruit trees within enclosed courtyards can be seen today in the timeless Nubian village on Elephantine Island at Aswan, Egypt where date palms grow today as they no doubt did in earlier millennia.

Other trees, not valued for their food value, were collected or acquired for the gardens of the Pharaohs by trade from throughout Egypt and adjoining countries. Carvings at Deir-el-Bahan show an expedition to Somalia; commissioned by Queen Hatshepsut (c. 1489–1460 BC), to collect and transport incense trees (*Boswellia carterii*) and other plants back to Egypt. These carvings document the interest on the part of rulers in the Ancient world to acquire trees of special value from considerable distances. Dixon (1969) states the Ramses III also imported incense and myrrh (*Commiphora myrrha)* trees from Arabia to have them planted in Memphis. Ownership of such trees demonstrated the wealth of the royalty and the power they could command in acquiring unique trees.

Trees were incorporated into the design of temples in ancient Egypt. The mortuary temple of Mentuhotep (c. 2020 BC) at Deir el-Babari opposite modern Luxor had an entrance way through a grove of trees (Wilkinson 1998). These trees were

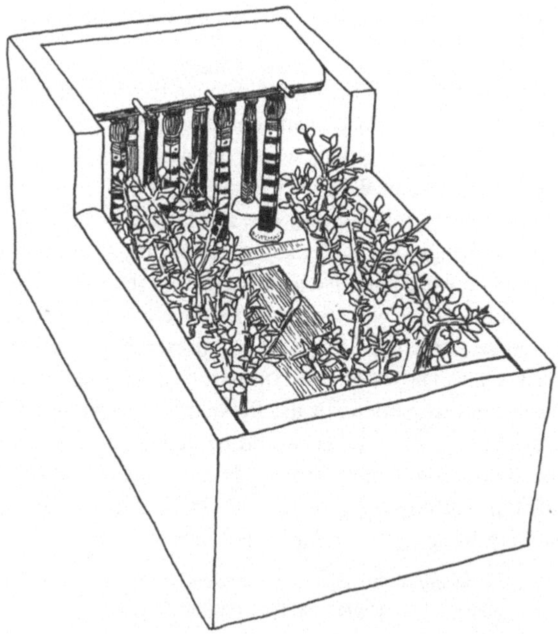

Fig. 4.6 Garden of Meketre c. 2050 BC (Drawn from exhibit at the Metropolitan Museum of Art, NY)

planted in large circular pits up to 30 feet deep in a grid pattern on either side of the ramp leading up to the temple. Remains of roots, bark, and leaves found in these pits indicate that sycomore and tamarisk had been planted. Although mortuary temples were erected outside of cities in ancient Egypt, temples to honor the Egyptian gods were often adjacent to or within cities. Trees planted at these temples not only provided shade and the definition of sacred space, they also served as a source of food for the priests. Temple gardens, vineyards, and orchards, often adjacent to the temples themselves served as sources of income for the support of temples and their priests.

Although trees were incorporated into urban areas of Ancient Egypt their presence was limited to palace gardens and the gardens of important government officials. Trees at temples were also part of these early urban forests when the temples were adjacent to or within the Egyptian cities. No evidence exists for street trees or for the equivalent of public parks where trees may have grown. The common people living in these cities had very limited outdoor space around their dwellings. This outdoor space was used for a variety of activities, but seldom was large enough for the planting of vegetables let alone trees. At the workmen's villages at Amarna and Mirgissa walled areas adjacent to the houses contained grids of square growing pits (about 12 inch by 12 inch by 6 inch deep) carved into the bedrock. These were filled with soil and used to grow vegetables (Griffith 1924). Fruit and nut crops consumed by the workmen were transported from the orchards along the Nile.

Trees in the Ancient Cities of Persia, Greece, and Rome

Cities in Persia and the Mediterranean basin initially used trees in a somewhat similar fashion to their use in ancient cities of Egypt, but later added new uses not found in Egypt. Palace gardens were common to the urban landscapes of ancient cities in the Mediterranean areas as well as in Persia. The palace gardens of the Pharaohs did not, however, resemble in size, tree species diversity, nor use the gardens of the Persian Kings. These Persian gardens are commonly referred to as 'paradise' gardens. The word paradise is derived from the Persian word *pairidaeza*, which means an enclosure. It was used in reference to the hunting parks of the Persian Kings. Paradise gardens of the Persian kings were luxurious beyond the means of other aristocrats. They were an expression of the great wealth, power, and extravagance. The paradise gardens of the Persian rulers extended beyond the confines of urban areas and in some cases would be considered ex-urban. In spite of their locations they had a significant influenced on the incorporation of trees into urban areas of the ancient world. Xenophon (c. 430–355 BC) complimented the paradise garden of the Persian King Cyrus at Sardis as follows:

> …the beauty of the trees in it, the accuracy of the spacing, the straightness of the rows, the regularity of the angles and the multitude of sweet scents.

The most famous example of the adaptation of a paradise garden into an urban area in pre-Greek and Roman world was the Hanging Gardens of Babylon. Nebuchadnezzar II (605–562 BC) is credited with having built these gardens for his Persian wife Amyhia, who longed for the mountain landscapes and meadows of her youth in what is now Kurdistan. The mountains of Kurdistan support a mixed deciduous forest that include oaks (*Quercus persia, Q. infectoria, Q, Libani*), ash (*Fraxinus rotundifolia, F. syrica*), maple (*Acer cinerascens*), pear (*Pyrus syrica, P. glabra*), pistacia (*Pistacia atlantica*), cotoneaster (*Cotoneaster nummalaria*), berberis (*Berberis integerrima*), oriental sycamore (*Platanus orientalis*), juniper (*Juniperus excelsa*), and honeysuckle (*Lonicera nummularifolia*) (Tabatabaie 2005; Mossadegh 2004). One could speculate that some these species were used in the gardens built for Amyhia.

The Persian paradise garden and its equivalents in the Middle East supported a more diverse mixture of tree species than the gardens of the Pharaohs and their administrators. Carroll (2003) indicates these gardens supported apple, cedar, fig, juniper, oak, palms, pear, pines, plane, pomegranate, olive, orange, terebinth, willow, and a variety of nut trees. The greater number of species was due to the more favorable climate for tree growth than occurred in Egypt, a more diverse native tree flora to draw from, and the desire to provide some habitat and cover for the animals that were released into these gardens for hunting. Persia was the center for the domestication of a number of species of *Prunus* that have since been used in orchards and gardens. One would have expected to find these in the paradise gardens of the Persian rulers.

The equivalent of the royal garden in Egypt and the paradise gardens of the rulers of Mesopotamian and Persia were not present in Greece during its golden age of democracy (fourth to fifth centuries BC). The practice of democracy prevented leaders from acquiring either the great wealth or the power necessary for the building of the great royal gardens seen in other parts of the ancient world. Gardening was, however, very important to both the rulers and common people of Greece. Greek towns were surrounded by greenbelts of vegetation primarily composed of garden plots and suburban orchards. These gardens were be watered from adjacent rivers. The scarcity of water in Athens and other Greek cities, prior to the introduction of lead water pipes during the Roman era, limited the possibility of gardening on any large scale within the cities (Carroll 2003). The defensive design of many of these early Greek cities would also dictate against the use of land for gardening within the city walls. It was, however, common for each house to have a designated garden plot outside of the city wall. These garden plots could be sold separately from the house and it was also possible for garden space to be rented. One can imagine that working in these garden plots and orchards was a welcome relief from the crowded conditions of the ancient Greek cities, as it is today for apartment dwellers in Moscow who tend their garden plots around the margins of the city.

In contrast to the absence of pleasure gardens during the fifth and fourth centuries BC in Greece, later Greek rulers admired the pleasure gardens they observed in their conquests of Persia and attempted to copy them on royal estates. This trend was more fully expanded in the rural villas built by Roman emperors and military leaders made wealthy by the conquests of foreign lands. Although the Roman villa was an exurban phenomenon, smaller versions of villas were built in Rome and as the city grew suburban villas were absorbed. Typical Roman villas supported a central garden around a rectangular pool somewhat reminiscent of the Egyptian garden, but with less use of trees. Porticoes provide the shade needed in the warm Mediterranean summers. Large areas of orchard trees were common to the design of the Roman villa. Vineyards and vegetable garden plots accompanied the orchards to provide a somewhat sustainable agricultural unit.

In China a tradition of palace gardens extends back to the early emperors, at least 2000 years (Thacker 1979). Although of independent origin and characterized by uniquely different design, these palace gardens served the same functions for the Chinese emperors as did paradise gardens for the early rulers in the Middle East. Like the Persian paradise gardens the early gardens of the Chinese Emperors were used for hunting grounds, keeping animals for show, and to produce flowers for domestic use (Cowell 1978). When the Chinese nobles acquired the wealth necessary to build and maintain their own gardens, the early Emperors decreed that they alone could grow certain trees, like the lacebark pine (*Pinus bungeana*), in their gardens (Wu 2004). The emperor's gardens also served the plant collecting interests of certain emperors. Emperor Wu Ti (c 140–86 BC) is credited with establishing the first botanical garden in Asia in 111BC in the Han capital of Ch'ang-an (Siren 1949). Marco Polo (c. 1254–1324) describe the famous 'Green Mountain' arboretum established by the Mongol Emperor, Kublai Khan, in the palace garden in Cambaluc [now Beijing] (Yule and Cordier 1992):

> Moreover on the north side of the palace, about a bow-shot off, there is a hill which has been made by art (from the earth dug out of the lake); it is a good 100 paces in height and a mile in compass. This hill is entirely covered with trees that never lose their leaves, but remain evergreen. And I assure you that wherever a beautiful tree may exist, and the Emperor gets news of it, he sends for it and has it transported bodily with all its roots and earth attached to them, and planted on that hill of his. No matter how big the tree may be, he gets it carried by his elephants; and in this way he has got together the most beautiful collection of trees in all the world.

No doubt a similar statement could have been made for the trees assembled in the gardens of Tenochtitlan and Iztapalapan by Montezuma and his predecessors.

Trees were used in public places primarily to designate sacred sites as cities evolved in Greek and Roman times. Processional routes to sacred sites were lined with trees to both provide shade and delineate the pathways. Pausanias, the second century AD Greek traveler and geographer, reported the presence of few trees within the many Greek cities he visited. His reports of trees are primarily limited to scared groves at the edge of cities or at sacred springs. Some of these he believed were planted while others were naturally occurring. He reported the presence of oaks, olives, bay-laurel, cypress, and plane tree in the different groves he described (Jones 1935).

Shade is an important element for outdoor comfort in the Mediterranean climate during the long dry summer. Greek and Roman architecture utilized shaded walkways along commercial streets to provide protection from summer sun and winter rain. Streets were narrow and no accommodation was provided for street trees. In the more open agoras, space was available for trees. It is probable that the trees in the ancient agoras were relict trees from the natural and agricultural landscapes in which these agora were established. One most likely would have encountered holm oak (*Quercus ilex*), Kermes oak, (*Quercus coccifera),* and olive (*Olea europaea).* In the fifth century BC the Athenian agora was planted with plane trees (*Platanus orientalis*) to provide a cool shady space for citizens to congregate (Clough 2001).

The first Greek philosophers often taught as they walked in the shade of colonnades in Athens and beneath trees in the agora. Later on educational institutions, called "gymnasia", grew up in the suburbs surrounding Athens. The Academy, Lykeion, and Kynosarges were gymnasia established among sacred groves outside of the city wall. These gymnasia incorporated natural woodlands along the Kephissos, Eridanos, and Ilissons rivers as parts of their developed sites (Fig. 4.7). Aristophanes recognized the elm, plane, poplar, and olive trees in his praise of the woodlands found at the Academy (Roche 2005). These were most likely the small-leaved elm (*Ulmus minor),* oriental plane (*Platanus orientalis),* black poplar (*Populus nigra*), and olive (*Olea europaea).*

Philosophers who taught at these gymnasia often built houses nearby and in some cases established their own schools. Theophratsus (c. 370–286 B.C.) owned a house with a colonnade and garden containing a scholar's walk near the Lyceum. It was a place for teaching his students and may have been the first botanical garden in the western world (Thacker 1979). Theophrastus willed his property to his students that they might continue to use it as a place of learning.

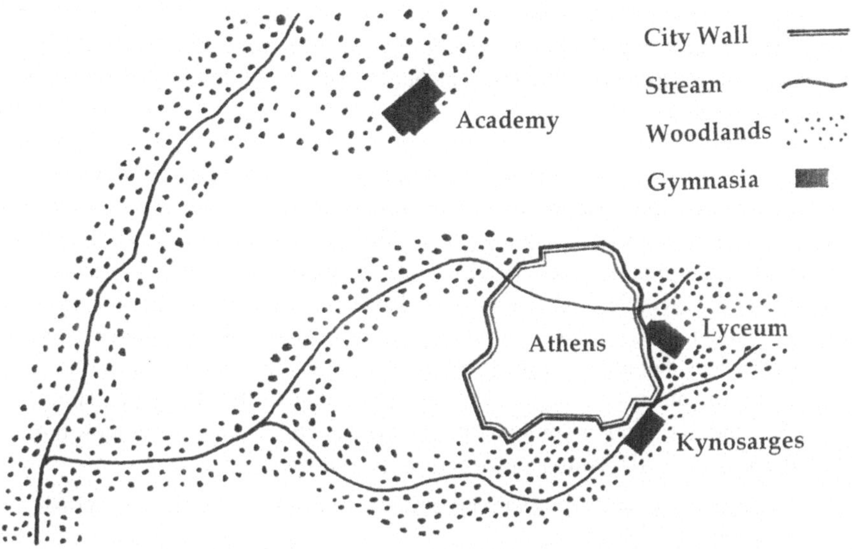

Fig. 4.7 Natural woodlands and sites of the gymnasia outside of Athens ca. 450 BC

The gymnasia also included athletic fields that were near the streams and adjacent to the woodlands to provide water and shade for competing athletes. The grounds surrounding these gymnasia with their athletic fields, woodlands and streams became defacto public parks during sporting events when the citizens of Athens came to observe the competition.

Naturally occurring sacred groves associated with spiritual places in the landscape of the ancient world inspired the religious gardens that were planted to honor Greek and Roman gods in urban areas. Certain gods were associated with vegetation, gardens, flowers, and fertility. These gods were honored with individual gardens or shrines within the private pleasure gardens of the leading citizens. Different tree species were associated with certain Greek gods: Zeus (oak), Apollo (laurel), Aphrodite (myrtle). Trees were often planted around new temples. The temple of Zeus at Nemea was reported by Pausanias to have had a grove of cypress trees. Excavations at the site have revealed twenty-three planting pits dug into the bedrock adjacent to the temple (Miller 1978). In other parts of the Mediterranean world trees were also planted adjacent to temples. The temple erected for the Assyrian god Ashur was surrounded by more than 2000 trees planted in pits dug into the bedrock (Carroll 2003).

Funerary groves were also planted by the Greeks to honor their dead. In reference to the funerary groves of heroes of an earlier period Homer in the Iliad made the following reference:

> He killed Eetion but did not strip his armor, for his heart respected the dead man, but burned the body in all its elaborate war gear and piled a grave mound over it, and the nymphs of the mountains, daughters of Zeus of the aegis, planted elm trees about it.

The grave mound with an adjacent grove of trees was proposed by Plato (2000) as the ideal mode of burial. However, the grave mound was replaced by monumental tombs and the simple grove by more elaborate gardens by the Hellenistic period (Fraser and Nicholas 1958).

During later Greek and Roman times pleasure gardens were planted in urban areas as well as at private villa for the enjoyment of individuals of considerable wealth. Although these gardens did not capture the opulence of the ‘paradise’ gardens of Babylon and Persia, they did provide the wealthy citizen with a refuge from the typical urban landscape of the Greek and Roman city. Small gardens for the production of vegetables and herbs were also maintained on the urban estates of wealthy Roman citizens. These utilitarian gardens lacked the adornments of the pleasure gardens, but no doubt provided great satisfaction to their owners. The vast majority of Roman citizens was not rich and was unable to afford gardens. They did, however, adorn the windows of their apartments with flowerpots and window boxes. Pliny the Elder (77 AD) commented on these as follows:

> The common people of Rome offer the eye a reflection of the country with their miniature gardens in their windows.

Rome also supported public gardens that were open to all citizens. Many of these were built by wealthy Roman aristocrats to gain recognition by the citizens of Rome. Pompey the Great had a grove, theater, basilica, and market built in the center of the city in 55 BC (Gleason 1994). Others were initially the private estates of leading citizens such as those of Julius Caesar during the Republic that were willed by their owners to the citizens of Rome. The establishment and maintenance of these gardens was facilitated by the availability of water in Rome. The Roman development of aqueducts and lead water pipes allowed for an economical watering of trees that had not been possible in Athens or the cities of Egypt. In the Empire period the number of public gardens increased. Grass and trees were planted around the *thermea* (social centers with public baths), libraries, and recreation grounds for the benefit of the general public. Pliny the Elder (77 AD) notes that in addition to booty and slaves, trees were “among the captives” in the triumphal processions of Roman generals returning from campaigns in the eastern Mediterranean and Asiatic provinces. Pliny specially mentions cherry, peach, apricot, and pistachio trees being among the “captives”. These were planted in the villas of the generals and enhanced the agriculture of the western portions of the Roman Empire.

The Roman villa was an exurban phenomenon serving as a place for the wealthy Romans to escape the discomforts of the city. Typical Roman villas supported a central garden around a rectangular pool somewhat reminiscent of the Egyptian garden, but with less use of trees. Porticoes provide the shade needed for the warm Mediterranean summers. Large areas of orchard trees were common to the design of the Roman villa. Vineyards and vegetable garden plots accompanied the orchards to provide a somewhat sustainable agricultural unit. Villas were sometimes absorbed into Roman cities as a result of urban expansion adding trees to the cities.

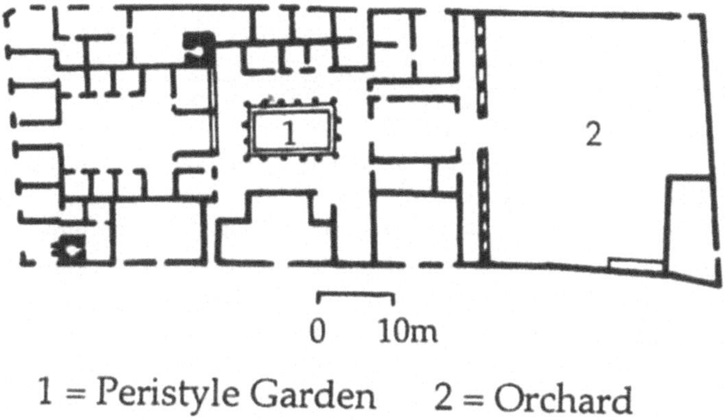

Fig. 4.8 Home of Pansa (c. 70 AD) (Redrawn from Carroll 2003)

The urban houses of well-to-do Romans provided sufficient space for the incorporation of plants. These were usually planted in a peristyle garden. The house of Pansa, in Pompeii (Fig. 4.8), had the traditional peristyle garden and a walled orchard that included vegetable beds at the back of the house (Jashemski 1979). Carefully pruned ornamental trees and fruit bearing trees were sometimes incorporated into the peristyle gardens of Roman houses. Commonly used trees included Italian cypress, fig, cherry, pear, and apple (Jashemski 1981). Rome was surrounded by a 'greenbelt' of large private estates, which from the first century onward were mostly under the control of the Emperor. Some of these were converted in part to public parks as the city expanded. Cowell (1978) suggests that this urban green belt was recognized for its role of maintaining air quality by the Roman Senate and debates were conducted over the feasibility of converting the villas and their adjacent agricultural lands to urban housing.

The incorporation of trees in the ancient cities of Egyptian, Babylonian, Persian, Greek, Roman, Chinese, and Mexico suggest a pattern of the use of trees that was copied or re-invented as cities evolved. These include (1) the use of trees to provided food and medicinal products, (2) the introduction of exotic species (often brought from considerable distances), (3) the use of trees for shade, (4) incorporation of trees public and private gardens for the pleasure of urban citizens, (5) tree planting to mark sacred sites and temples, (6) the creation of botanical gardens and arboreta, and (7) the maintenance of natural woodlands and orchards in urban green belts. These uses of trees in urban areas are not unlike those found today in metropolitan areas throughout the world. However, much of the knowledge of the use of trees in urban areas in the Ancient world was lost in Europe during the Dark Ages following the fall of Rome. Centuries were required to regain this knowledge, a level of urban security, and the economic means needed to replicate the variety of uses of trees seen in the cities of the Ancient World.

The Middle Ages in Europe

The Western Roman Empire collapsed via a series of foreign invasions following the revolt of the Goths in 398 AD. The Goths first plundered Greece and then invaded Italy. In 410 AD Rome was captured by Alaric. The successful challenge to Roman military authority by the Goths set the stage for subsequent invasions of barbarians coming out of the east. Cities in Greece, Italy, northern Europe and north Africa were pillaged and often burned by successive waves of Goths, Visigoths, Vandals, and Huns. Urban populations were often compelled to consume trees in both public and private gardens as a source of fuel during extended periods of siege. Invading armies often cut down or burned trees within palace grounds and at sacred sites to demonstrate their power over the defeated rulers and their gods. The great wealth of the Roman cities, which had supported private gardens and the maintenance of trees in public places, was destroyed. No doubt some forms of subsistence gardening were attempted in the open space areas of these war-ravaged cities. The experience of using public parks for vegetable gardens in European cities during World War II was akin to what probably took place in the besieged cities of the Roman empire (Stilgenbauer and McBride 2010).

The Dark Ages descended upon Europe with the fall of the Western Roman Empire. Nothing was written during this 300-year period (ca. 476 to 795 AD) to provide us with a view of urban vegetation (Thacker 1979). It was a time of political instability, spreading ignorance, and great suffering. Urban populations declined and the stability and wealth necessary for the establishment and maintenance of trees in urban areas declined. Although cities did not disappear after the fall of the Roman Empire, the sporadic siege of cities, famine, and plague took their tools on urban populations. The population of Rome dropped from several hundred thousand at the beginning of the fourth century AD to 30,000 by the end of the sixth century (Brown 1988). The reduction in urban populations and the increased necessity for the defense of cities had a significant impact on nature of cities. The grid of streets and outdoor public spaces common to many Roman cities were compromised by the development of defensive enclaves within the cities. Extended sieges of cities necessitated the felling of trees to provide fuel and materials for building defensive bulwarks. Densification of urban populations in Roman cities often came at the expense of open space that had previously supported trees.

Many cities survived the ravages of the Darks Ages and persisted as population centers through out the Middle Ages. Their survival initially depended upon defensive structures, which provided refuge from marauding bands. The old Roman amphitheaters at Arles and Mimes, in Provence, were transformed by town people into fortified centers by closing the entrances and using their heavy masonry walls as ramparts (Fig. 4.9). Two thousand people were able to survive in the converted amphitheater at Mimes. In other cities, portions of the defensive walls build in Roman times were restored to provide security. These town walls, like the defensive walls of the later Medieval castle, provided protected areas in which some cultivation of plants could take place. Cultivation also occurred outside of the city walls and attempts were made to maintain orchards.

Fig. 4.9 Amphitheater in Arles converted to fortified housing during the Dark Ages (Redrawn from Guibert, British Museum)

During the Dark Ages the Roman Catholic Church slowly gained control over much of Europe through assuming administrative control of cities and through the establishment of monasteries in rural areas. A monastery functioned as small cities and often attracted peasant populations that lived outside of the monastery walls. The peasants served as labor force for certain monasteries whose monks were devoted to copying of documents, prayer, or other religious tasks.

The monasteries preserved some features of the villa and urban gardens of the Roman Empire. In time they provided the germplasm of ideas and plant materials from which urban vegetation re-emerged in cities during the Middle Ages. Our knowledge of the early monastic gardens is limited to a few documents coming out of the Dark Ages. One of the most remarkable of these is the plan for a monastery and its associated buildings, which was preserved in the monastery of St. Gall in Switzerland (Horn and Born 1979). This plan shows three gardens each with a separate purpose within the outer walls of the monastery (Fig. 4.10). An herb or physic garden located next to the infirmary, a vegetable garden, and an orchard that was also to serve as a grave yard for the departed monks. A similar orchard/cemetery was depicted in Prior Wibert's plan (c. 1165) for a new water system for the Priory of Christ Church in Canterbury.

The plan for St. Gall includes a cloister but makes no reference to any trees within the cloister. Descriptions and illustrations of monastic cloisters from later in the Middle Ages indicate that cloisters were commonly divided into four areas by two intersecting paths and a tree or fountain stood in the middle. When trees were used, species of pines and juniper were favored trees because of they symbolized everlasting life (Landsberg 1995). The open areas were planted with turf and every

Fig. 4.10 Orchard and cemetery from the plan for St. Gall (Redrawn from the Plan for St. Gall, Horn and Born 1979)

effort was made to keep them as green as possible. The color green provided by the plants in the cloister was believed to be of importance to the well being of the monks and nuns confined to monasteries and nunneries. In the twelfth century AD Hugh of Fouilloy wrote (Eco 1986):

> The green turf which is in the middle of the cloister refreshes encloistered eyes and their desire to study returns. It is truly the nature of the color green that it nourishes the eyes and preserves their vision

Although monastic gardens were utilitarian in nature, they also contained sections or even separate gardens where the sacrist raised lilies, roses, and other flowers used to adorn alters and shrines and to provide crowns and garlands occasionally worn by priest when performing services. Foliage and branches of yew, hazel, bay and holly were used on certain occasions and may have been collected from trees in the cemetery/orchards or from trees included in the sacrist's gardens (Cowell 1978).

The hard work, but ultimate satisfaction of planting and maintaining a monastic garden, was extolled in Abbot Walafrid Strabo's (c 809–849) poem *Liber de cultura hortorum* (Payne 1966). The techniques of gardening and tree care described in Abbot Walafrid's poem had their roots in the art of gardening developed in Greek and Roman times. The survival of these gardening skills in the medieval monasteries was fundamental to the establishment and maintenance of trees in urban areas in the later Middle Ages.

Ingersoll (1995) credits the monastery as an important institution and form for the evolution of European cities. He contends that, "they maintained the rational typologies of an earlier urban order, that had lapsed during the devastation of the cities" and, "they often gave rise to a new city or provided pockets of order as new additions to cities."

The domination of Western Europe by Charlemagne in the later part of the eighth century AD led to a brief period of political stability and security, which contributed to the rejuvenation of cities. Charlemagne's court issued regulations for the administration of cities within the empire (*Capitulare de villis vel curtis impreii,* 795 AD). These regulations list some 73 plants and fruit trees which must be grown for the public good. The plants on this list were to be grown to provide herbs, vegetables, fruits and nuts. The list is a compendium of the plants commonly raised in monastic gardens of the time. Included on the list were apples, pears, chestnut, peaches, quince, hazelnut, almond, mulberry, bay, stone pine, fig, walnut, and cherries (von Fischer-Benzon 1894).

A century after the collapse of Charlemagne's empire much of Western Europe was fragmented into feudal estates. The typical estate was comprised of about 1000 acres of land that would have been divided into about 300 farms. The estates were run by vassals who owed tribute and military service to regional lords. The peasants who farmed the land were typically required to give half of their crops to the vassal and were not allowed to leave the estates on which they worked. The vassal provided security for the peasants to protect them from marauders. Castles evolved as a means of providing protection to local populations and secure housing for the local vassal. Initially the castle was simply a bare stronghold without space for gardens or trees. The open space within the castle walls served to accommodate the surrounding population and war horses during periods of siege. These early castles consisted of single or double enclosing walls, in the center of which was a massive tower that served as the center of defense and the feudal lords dwelling. Plots of land outside of the walls were used for growing vegetables and orchard crops for the household of the vassal.

Castles passed through various stages of development in the Middle Ages (Clark 1884). A major revolution, of consequence to the inclusion of gardens and trees within castle walls, took place in the construction of castles in the thirteenth century. Castles were expanded and additional structures built within the protective walls. Their quality as fortresses began to embrace that of a residence suitable to the life of a lord living in the middle of his domain surrounded by a small court and a garrison. Clark (1884) suggests that they became "rather palace-castles than castle-palaces". This revolution in design provided space for pleasure gardens within the castle walls. It also brought about an extension of defensive walls and motes to enclose orchards and vegetable plots which formerly were maintained outside of the castle walls.

The excellent compendium of illustration and descriptions of medieval gardens assemble by Crisp (1924) shows the basic character of the pleasure gardens found within castle walls. These gardens were limited in area, generally square, and surrounded by fences or walls to provide a sense of enclosure (Fig. 4.11) Trees were used sparingly to accentuate particular points in the garden and to provide shade. Harvey (1989) lists five trees that were commonly used in medieval gardens in the early part of the sixteenth century: almond, bay, peach, pine, and plum. Living ground cover used in medieval gardens was either grass or a "flowery meade" (a recreation of a meadow rich in flowers). Grass and flowers were also planted within

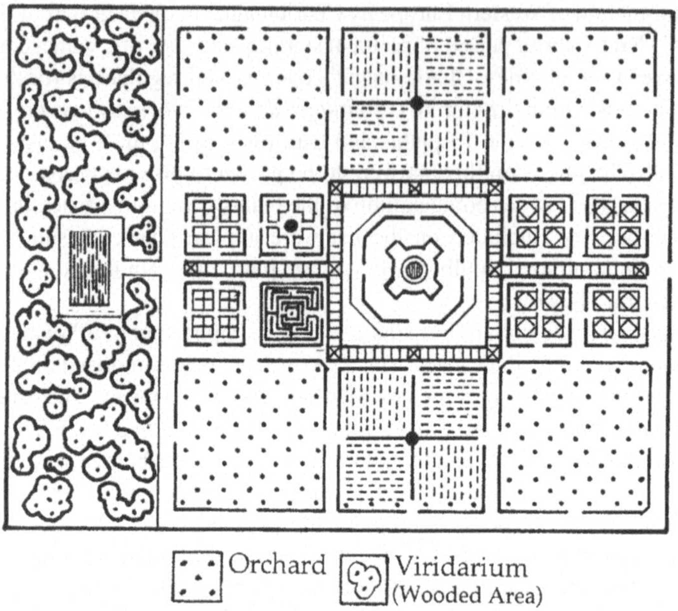

Fig. 4.11 Reconstruction of the largest garden described by Pietro de Crescenzi (c. 1300 AD) (Redrawn from Pizzoni. F. 1999)

the orchards that were "brought" into the castles by the extension of walls and motes in the thirteenth century. Raised beds, in which flowers were planted, were a common feature of castle gardens as were turf seats. Seats were commonly made around the base of a tree by building a stonewall or wattle to hold soil upon which was laid turf (Landsberg 1996). In the fourteenth century grass-covered mounts, artificial mounds of earth, appeared first in the castle gardens of France. These mounts provided a view over the castle walls to the surrounding countryside. They probably served both as a defensive lookout and a vantage point for the pleasure of looking out from the confining walls of the castle.

A number of factors began to come together in the tenth century that led to the rejuvenation of the cities and the establishment of new towns. Mumford (1961) suggests that a combination of military protection, economic needs, increasing populations, security for markets, the revival of trade, coinage of money, and the freedom from feudal lords was responsible for this renewal of urban living. New town charters were granted and colonists built towns to extend the defenses of feudal barons or the exploit the agricultural potential of the landscape. In parts of northern Europe and the British Isles forests were cleared for agriculture and new towns. This pattern of conversion of forested landscapes to new towns and their surrounding agricultural lands was carried to North America in the seventeenth century. The new towns established in the Middle Ages in Europe often made space available for individual herb and vegetable plots and for orchard trees. Orchard trees were sometimes planted at a low density to create a *sub arboribus*, or small shaded pasture for

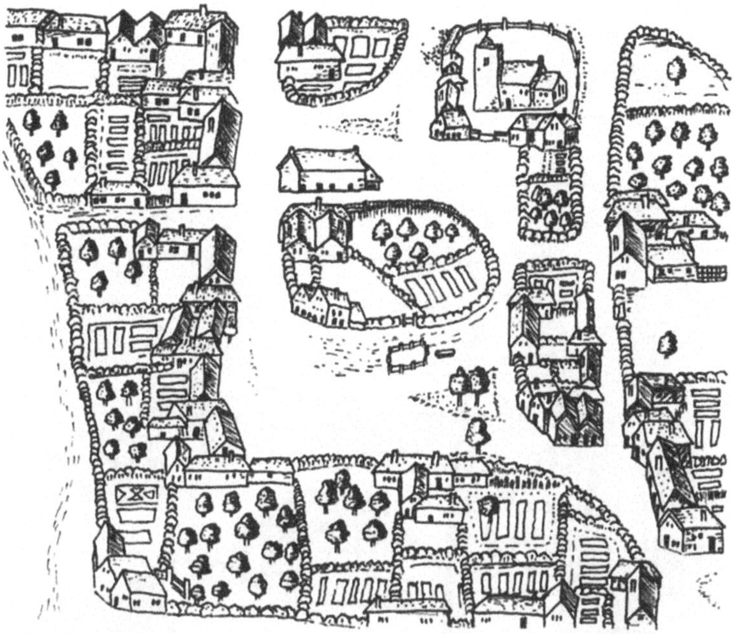

Fig. 4.12 Wilton (c. 1565) (Redrawn from Wiltonshire Records Office: Wilton Estate, Survey of Lands of First Earl of Pembroke, 2057 SB)

milk cows kept within the city walls (Landsberg 1998) (Fig. 4.12). A plan of the Tutor city of Hull, believed to be drawn around 1538 to illustrate the city fortifications (de Boer 1973), shows a number of plots between the city wall and the houses that were either orchards or shaded pastures. A mid sixteenth century plan for the English town of Wilton show the inclusion of garden plots, orchards, and shaded pastures adjacent to houses in peasant closes (Landsberg 1998).

Jackson (1984) described the "movable" nature of the peasant house during the Middle Ages. The typical house could easily be disassembled and the important structural members moved to another site where the house could be re-assembled without the need for a foundation. This enabled the placement of houses among trees as new villages were created. Such placement took advantage of the shade value of the trees. Evidence of the placement of houses among trees can be seen in the paintings of Peter Brughal (Fig. 4.13). Although Bruehel's paintings are from the sixteenth century, they depict peasant houses typical of the "movable" dwellings late middle ages discussed by Jackson.

The long sweep of the Middle Ages encompassed the near extinction of trees from urban areas to their re-incorporation in the medieval garden of the nobles and utilitarian gardens, orchards, and shaded pasture plots of the common people. Trees served the same functional, amenity, and spiritual roles as they had in cities of the ancient world, but at a much reduced scale and with a far fewer species. The notable exceptions to this were the cities in Spain under Islamic rule from the eighth to the

Fig. 4.13 Trees and peasant houses (Redrawn from the Numbering at Bethlehem (P. Bruegel, 1566, Musee Royaux des Beaux-Arts, Brussels)

fifteenth centuries. These cities supported extensive gardens that reflected in their design the paradise gardens of the Middle East. Many of the new cities established by the Muslims were surrounded by planted palm groves as can been observed today in the city of Elche in Spain.

The Renaissance

The Renaissance did not emerge suddenly out of the Middle Ages but slowly out of the urban setting and sophisticated intellectual environment of medieval society. It had its roots in the Crusades of the eleventh century when Italian coastal towns took the lead in the reconquest of the Mediterranean and gained a level of independence from Papal control. Other roots go back to the "Carolingian Renaissance" of the late eighth and ninth centuries that saved many ancient works from destruction or oblivion. The revival of Roman law, Latin poetry, and Greek science in Italy and the emergence of commerce based on cloth in Flanders in the twelfth century also laid the seeds for the Renaissance that bloomed in Italy from the fourteenth to the sixteenth century. Much has been written by the historians in the nineteenth and twentieth centuries to define and understand the origins and the essence of changes that took place in Italy and the rest of Europe during the Renaissance (Michelet 1855; Burchardt 1877; Sapori 1940; Ferguson 1948; Helton 1964). A common thread coming out of this body of work is the emergence of the autonomous city-state and a new culture of laymen who sought models for society in the writings of the Greek and Roman philosophers.

The creation of the first Renaissance garden was an outgrowth of the renewed interest in classical texts. By the middle of the fifteenth century the typology of the Italian Renaissance garden and its associated villa had become established along the lines suggested in the treatises of the Roman period. Volume seven of Albertti's ten volume work on architecture prescribed the basic design of the Italian Renaissance garden (Alberti 1485). Such classical texts as those of Vitruvius (first century BC) and Pliny the Elder (first century AD) were fundamental points of reference for these gardens (Pizzoni 1999). Basic to the design proposed by Alberti were sets of terraced gardens that could be categorized as either (1) herbaceous/flower gardens, (2) orchards, and (3) bosques (tree groves). Alberti stressed that the gardens be divided into compartments that were to be edged with low hedges. Sweeping views from the more formal parts of the garden over the orchard plantings and the bosques were to be designed to provide visual access to the surrounding landscape. Alberti emphasized the need for tree pruning to maintain sight lines and views over the tops of trees.

A detailed analysis of the Renaissance garden is beyond the scope of this book because of their exurban locations and the considerable variation in design in different parts of Europe. However, two characteristic design features of the Renaissance garden should be pointed out because of their influence on the incorporation of trees in cities. These are the bosque and the tree-lined promenade. Excellent reviews of the Renaissance garden may be found in the works of Lazzaro (1990); Mosser and Teyssot (1991); Shepherd and Jellicoe (1953); Woods (1996); Pizzoni (1999).

Bosques were generally planted to represent the native woodlands of the area in which the garden was built. As such they were often outside of the walls defining the more formal portions of the Renaissance garden. Bosques were also used to frame water features in the garden and to surround a piece of sculpture or fountain in order to achieve a sense of discovery or surprise on the part of the people walking through the garden. In some gardens, bosques were also planted around small lawns and pruned into hedge-like walls to create "green rooms" within the larger garden. Trees in larger bosques were usually planted in a square grid, but sometimes a planting pattern known as a quincunx was employed (Fig. 4.14). A quincunx is a pattern where five trees are grouped to create a more complicated arrangement. This planting pattern was probably more common to orchards in the Renaissance garden that to the bosques. Sullivan (2002) has suggested that bosques were an important feature of the Renaissance garden for role in creating comfortable microclimates. They provided shaded areas for protection from the Mediterranean sun as well as a refuge from the wind. Some Italian Renaissance gardens, such as the Villa Doria Pamphill, Rome, had bosques of Italian stone pine referred to as "pineta". These were planted on a tight grid usually ten to thirty feet on center and the were trees pruned to pro-

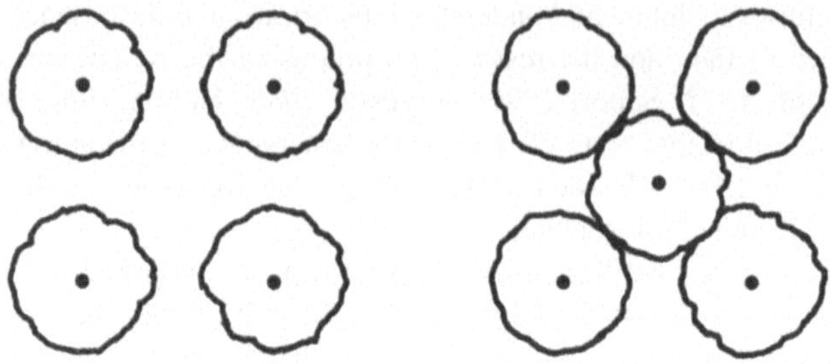

Fig. 4.14 Square grid and quincunx

duce a green umbrella like canopy. The bosque of the Renaissance gardens emerged in urban landscapes following the Renaissance when large public park were established in Europe.

A variety of tree species were used in the bosques of the Renaissance garden. This variation was due to differences in climate that allowed the use of more frost sensitive trees in the south of Europe and required the use of frost hardy species in the north (Lazzaro 1990). Palms could be employed in the warmer climates while chestnut trees served well in the colder regions. Variations in site conditions within a given garden also dictated the tree species that were used: willows and alders on wet sites, pines and oaks on drier sites. A list of 47 trees used in bosques of Italian Renaissance gardens during the fifteenth and sixteenth centuries has been complied by Lazzaro (1990) from various primary sources. She also lists an additional twenty-two orchard trees planted during the same period. The more commonly used bosque trees in the Italian Renaissance gardens in the fifteenth and sixteenth centuries were chestnut, cypress, elm, fir, holly oak, holm oak, juniper, laurel, maple, pine, and plane tree. The Renaissance garden was also used to display exotic trees much in the same way they had been displayed in the pleasure gardens of Persia and the gardens of the Pharaohs, Emperors, and Kings of the ancient world. The garden of Lorenzo de Medici at Careggi displayed an African ebony tree (*Diospyros sp.*) while Cosimo de Medici's garden at Castello exhibited a banyan tree (*Ficus benghalensis*) from India (Heikamp 1967).

The exploration of the western hemisphere and Pacific basin during the Renaissance resulted in the European discovery of new tree species. These trees were brought back to Europe like other treasures from the new world. Although new tree species were counted in the spoils of war during ancient times, it was not until the Renaissance that efforts were made by the first "plant hunters" to identify potential trees for landscape gardening (Whittle 1997). Early European botanists such as John Tradescants in England and Carolus Linnaeus in Sweden stimulated plant exploration for the purpose of science, but many of their plant hunters returned with trees that were to become important components of burgeoning urban forests in the late Renaissance. Tradescants is credited with the introduction of American sycamore (*Platanus occidentalis)* to England as a result of his collections in the Virginia

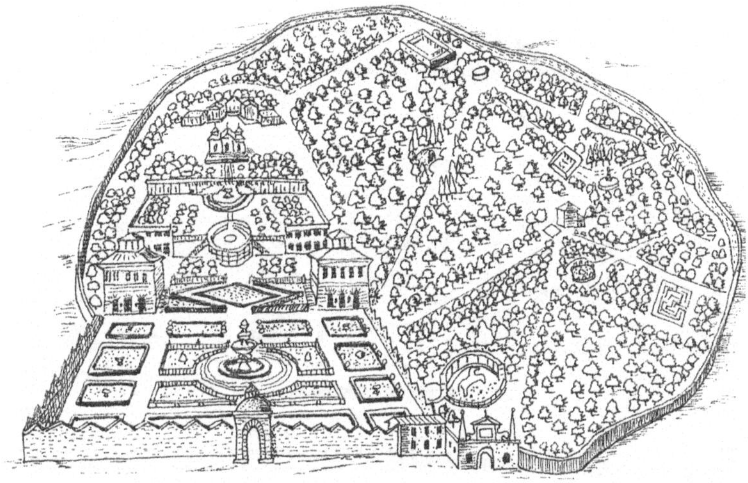

Fig. 4.15 Villa Lante (c. 1588) (Redrawn Lauro, 1612)

colony in the mid-seventeenth century (Spongberg 1990). His introduced specimen crossed with Spanish sycamore (*Platanus hispanica*) in Tradescant's garden to create a hybrid sycamore know as London plane tree (*Platanus x hispanica*). Thacker (1979) listed eleven trees that were introduced in to England between 1500 and 1700 for use in Renaissance gardens. Many of these (e.g., bald cypress – *Taxodium distichum)* were no doubt imported to enhance the reputation and status of the owners of the gardens; however, many of these trees subsequently became important street and boulevard trees, such as horse chestnut (*Aesculus hippocastanum*).

In the second half of the sixteenth century citrus trees, especially lemon, became very popular in the orchards of the Italian Renaissance gardens (Pizzoni 1999). The susceptibility of citrus to frost damage required the construction of orangeries (green house like structures where citrus trees could be protected in the winter) in the Renaissance gardens of northern Europe. During the summer the citrus trees were carried from the orangeries and placed in the garden. An interesting example of an orangery can be found in London's Kensington Garden, unfortunately it is no longer used to maintain citrus trees during the winter.

Tree -lined promenades were common in Renaissance gardens. The Renaissance garden was designed for groups of people to walk about and enjoy the features of the garden. Trees lined the many walkways either in single or double rows. They were also often planted in the hedges along the walks. Their presence served to reinforce sight lines and to create more pleasant microclimates. Examples of the tree lined promenades can be seen in the bosque at Villa Lante (Fig. 4.15) and in the orchards of Villa Medici, Castello (Fig. 4.16). Elms were used along the avenues crossing the bosque at Villa Lante and along some orchards in the more formal part of the garden (Rome 1588). In the orchards at Villa Medici, Castello cypress trees were planted along the walk ways and along the long canal designed to carry water

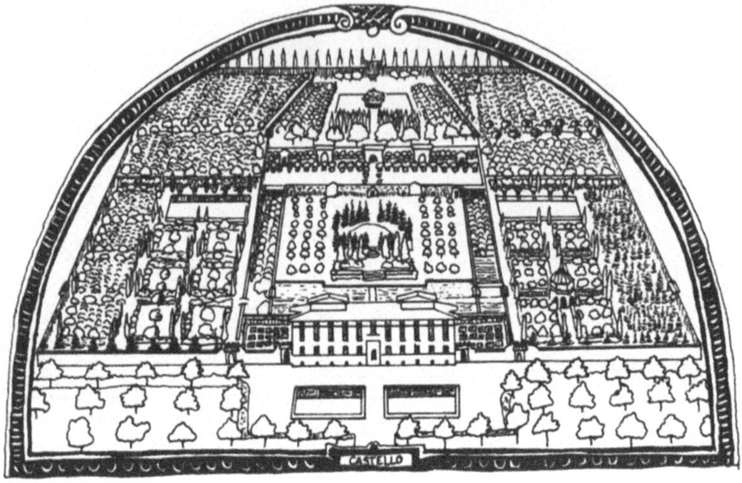

Fig. 4.16 Villa Medici, Castello (Villa Della Petraia) c. 1600 (Redrawn from Villa Della Petraia by Guisto Utens, from Pizzoni 1999)

to the western portion of the garden. The garden at Villa Montalto was also noted for the use of cypress along the avenues throughout the garden. Other trees used to line avenues in Italian Renaissance gardens included plane, fig, oak, olive, fir, chestnut, and various fruit trees (Lazzaro 1990).

Although trees had been used to line the approach to sacred sites in Egypt and Greece, and to provide shaded walks at the Greek gymnasia, their use in the Renaissance garden far exceeded these earlier uses both in the scale of the avenues they lined and in the variety of species used. As the Renaissance bosque served as a model for groves of trees planted in urban parks in the eighteenth and nineteenth centuries, so too did the tree-lined avenues in these Renaissance gardens provide a template for the tree-lined avenues and streets so familiar to today's urban landscape.

The introduction of tree-lined avenues in Europe occurred in the Renaissance as a result of the need to construct new defensive walls that could withstand cannon fire and protect urban populations growing up beyond the existing walls of many cities. Former city walls that were no longer of defensive value became popular places for people to stroll. The height of these walls gave a prospect to the countryside and often a chance to breathe fresh air. Obsolete city walls were initially planted with trees in Antwerp and Lucca in the 1570s to provide promenades for public use (Fig. 4.17) (Girouard 1985). This pattern spread across much of Europe in the seventeenth century. In the 1670s Louis XIV ordered the planting of trees atop the remains of the city wall of Paris in the vicinity of Place du Martin. Peter the Great converted the former walls around Moscow into tree- lined ring roads in the early part of the eighteenth century. In the twentieth century the Chinese city of Hefei (Anghai Province) converted its centuries-old, city wall into a tree-lined avenues and a circular park. A remnants of tree lined promenades planted on city wall can be

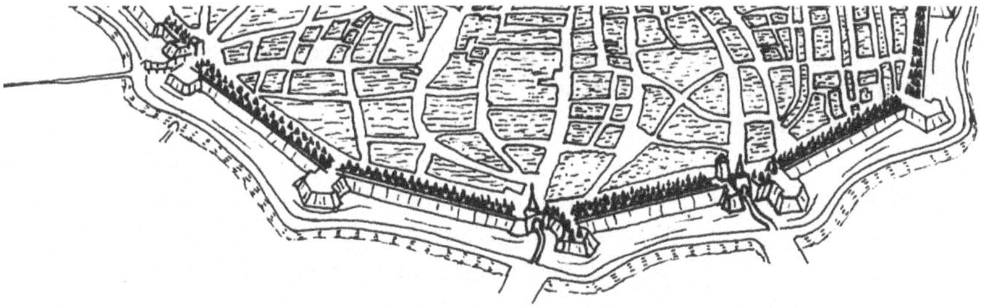

Fig. 4.17 Trees planted on the Antwerp city wall (c. 1635) (Redrawn from Girouard 1985)

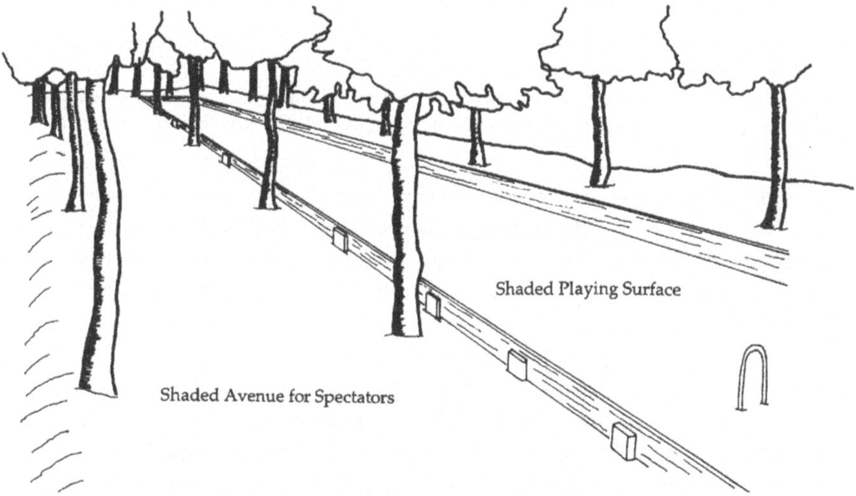

Fig. 4.18 Court for the game of pall mall (Redrawn from Girouard 1985)

seen in the Italian city of Ferrana along a portion of Via IV November (Jacobs et al. 2002). Modern tree lined promenades can be enjoyed along the restored sections of the city wall in Dresden.

In addition to the planting of promenades along the tops of obsolete city walls, trees were planted along the margins of rectangular spaces, known as malls, used for a game called "pall mall" that was somewhat similar to croquet (Fig. 4.18). The trees provided shade both for the players and the spectators. Two malls were constructed in Paris in the 1590s (Lawrence 1995). Illustrations of these early malls show a row of trees on either side of the mall surface upon which the game was played and a parallel row to create a shaded walkway along the playing area. In London, Pall Mall was in use by 1650. In that same year Berlin had a mall know as Unter den Linden for the linden trees (*Tilia cordata*) that were planted adjacent to the playing surface. It was subsequently converted into a tree-lined street and became one of the most fashionable streets in Berlin. Throngs of promenadors were drawn to the tree lined malls in Paris, London, and Berlin to enjoy casual conversation as the strolled beneath the trees. This activity was so popular that people in London referred to the hour when the greatest numbers of people were promenading as "High Mall" (Garreau 1991).

Trees were planted along canals in Holland starting in the early seventeenth century (Burke 1960). This type of planting brought trees into the centers of Dutch towns in contrast to the more peripheral planting of trees on old city walls. Tree-lined canals were reported by Parkinson (1784) in the Dutch colonial city of Batavia (now Jakarta) when Captain Cook landed the Endeavor there for repairs in 1770.

Trees had been planted along the major approach roads to French towns in the sixteenth century. These roads, such as the Champs Elysees in Paris, were later engulfed by urban growth to become major avenues. Tree planting along roads in China was reported by Marco Polo in the thirteenth century (Yule-Cordier 1992):

> The emperor moreover hath taken order that all the highways traveled by his messengers and the people generally should be planted with rows of great trees a few paces apart; and thus these trees are visible a long way off; and no one can miss the way by day or night. Even the roads through uninhabited tracts are thus planted, and it is the greatest possible solace to travelers.

In a similar fashion trees are believed to have been planted along roads in Japan as early as the third century AD. However, the first record (a public construction order) of county road tree planting as a government policy was in 759 AD (Cheng et al. 1999). Certain roads in India are believed to have been planted at least 1000 years ago. As in Europe, this practice often was followed by the absorption of these roads and their trees by the expansion of urban centers.

A new recreational pastime, carriage riding, evolved in Europe as security became more assured during the Renaissance. Paris adopted the idea of a special promenade (or cours) for carriages from the Casine Gardens outside of Florence. The Cours la Reine was laid out along the right bank of the Seine in 1616 and a similar cours was constructed along Rue St. Antoine a few years later (Lavendan 1975). Trees were planted in four rows to form a 60 foot-wide central drive for carriages, and two adjacent 20 foot wide sidewalks for pedestrians (Fig. 4.19). The success of the cours in Paris resulted in a proposal to extend the cours completely around the city. In Bullet and Blondel's plan of Paris, published in 1676, a location for this cours was proposed (Girouard 1985).

The planting of trees in Renaissance cities, especially in Paris, on obsolete city walls, to provide shade for the game of pall mall, and to create elegant carriage promenades gave birth to a new urban terminology that referenced tree lined passageways. Some of these terms like allee and avenue were adopted from Renaissance gardens, while others were of direct urban origin. The term "mall" designated the long rectangular pall mall court and its adjacent trees. Boulevard was derived from the French word "boulevard" meaning a large bastion. The practice of planting trees on old city walls, noted for their bastions, gave rise in the eighteenth century to the term boulevard (Girouard 1985). Promenade is a French word whose original meaning was "to take a walk in public". The place of the public walk and carriage ride assumed the name of the activity.

Fig. 4.19 Cours la Reine, Paris (Redrawn from an engraving by Aveline (Bibliotheque National Paris from in Girouard 1985)

Changes in the form of cities occurred during the Renaissance. City expansion was required to accommodate growing populations. New towns were established to accommodate increasing populations, to gain political and economic control over certain parts of Europe, and to accommodate the colonization of the New World. Urban populations in Europe grew rapidly after the middle of the fifteenth century. Antwerp doubled in population in the second half of the fifteenth century when its population neared 100,000. The increasing population led to a northerly expansion of Antwerp into an area that became known as 'New Town'. Existing public spaces and some streets within the old city were planted with trees to create public walks and gardens. Among these was the church yard of the cathedral, which was cleared of tombs and planted with trees sometime prior to 1569 (Girouard 1985). The Antwerp town council issued an order in 1578 for a triple row of trees to be planted on top of the old medieval city wall (Antwerp City Archives 1578) (Fig. 4.20). The greening of the city did not go unnoticed by foreign merchants and visitors to Antwerp. John Evelyn, the seventeenth century English arborist, upon visiting the city in 1641 wrote (Evelyn 1641):

> There was nothing about this city which more ravished me than those delicious shades and walls of stately trees, which render the fortified works of the town one of the sweetest places in Europe.

The impression Antwerp made on its foreign visitors stimulated an interest in the incorporation of trees into other European cities.

The increase in human population in the second half fifteenth century and the expansion of commerce in the sixteenth century also brought about the expansion of existing towns and the establishment of new towns both in Europe and the western hemisphere. It has been proposed that the designs for these new towns in Italy bor-

Fig. 4.20 Trees along the Antwerp city wall (Redrawn from Denis van Alstoot, 1620, Atle Pinakothek, Munich in Girouard 1985)

Fig. 4.21 Portion of Paignton (c. 1565) showing trees in gardens and along streets (Redrawn from Wiltshire Records Office, Wilton Estate, Survey of Lands of First Earl of Pembroke, 2057 SB)

rowed from the parterre (layout of flower beds) of the Renaissance garden (Tongiorgi 1983). However, a stronger case can be made for the adoption of concepts of new town planning from Roman times than from the arrangement of flowerbeds in Renaissance gardens (Reps 1969; Hiorns 1956; Zucker 1959). In reconfiguring existing towns and their expansion, the grid system was often employed for the laying out of streets. Radial street patterns were also employed in some Italian new towns. There is no evidence from existing plans for new towns in Europe that tree-lined promenades, avenues, or public open spaces were to be planted with trees. However, in the new sections of existing towns and in new towns, residential lots were usually of sufficient size to provide garden and orchard space for each residence (Fig. 4.21).

The plans for new towns during the Renaissance in the western hemisphere included a centralized open space and generally used a grid design for the pattern of street (Reps 1969). Town planning for Spanish colonial towns followed the "Laws of the Indies"(1573) which directed, in a chapter entitled "Royal Ordinances for the Laying Out of New Cities, Towns, and Villages", that new towns be laid out on a grid with a plaza (Nuttall 1921). The plaza was to be centralized except on coastal cities where it was to be open to the port. Although there is some confusion in the ordinances of the "Laws of the Indies" as to the location of the temple (or cathedral) in respect to the plaza (Gakenheimer 1959), most Spanish colonial cities were laid out with the temple on the plaza (Crouch et al. 1982). Administration building for civic affairs, town-council house, arsenal, and customhouse, were also designated for the plaza. Ordinance #113 designates that plazas were to be no less than 200 feet wide and 300 feet long, but not larger than 550 feet by 800 feet. Regardless of size, the length of the plaza was to be 1 ½ time its width. The ordinance suggested that an optimum size would be 400 by 600 feet. The size and shape was to accommodate fiestas in which horses were used and to provide a place to gather horses and other types of livestock should the town come under attack. The buildings facing the plaza and the streets entering the plaza were to have porticoes to provide shade and protection from the rain. The intended uses of the plaza and the shade provided by the porticoes dictated against tree planting.

The Laws of the Indies (Ordinance 3113) recommended narrow streets in warmer locations and wider streets in colder locations. Narrower streets in warmer climates utilized the shade of the buildings to reduce radiation on the streets and the adjacent buildings. In this case the shading function of trees was not required, and little space was available for the planting of trees along these narrow streets. In cooler climates the wider streets allowed sunlight to fall both on the street and the walls of the houses. The placement of trees along these streets would have resulted in unwanted shade in these colder landscapes.

Descriptions and early photographs of Santa Fe, New Mexico, a Spanish colonial town established under the "Laws of the Indies", verify the absence of trees in the plaza and as well as along the streets (Bancroft 1889; Pinos 1942; Crouch et al. 1982). One could argue that the absence of trees in Spanish colonial Santa Fe was due to the lack of water in its semi-arid environment; however, irrigation canals were early features of the town. They provided irrigation for vegetables raised in garden plots adjacent to the houses and could have served as a source of water for trees. J. B. Jackson has convincingly argued that a tradition of vernacular urban landscape design in the American southwest, with roots in Moorish Spain, did not include tree planting in plazas nor along streets.

Although plazas and streets were not planted with trees in colonial Spanish towns during the Renaissance, orchard trees were introduced into the individual garden spaces allotted for each dwelling and were also planted in small plots used as pastures for family milk cows in regions where summer rainfall or irrigation could be used to support the trees. A plan for the Spanish settlement of St. Augustine (Roberts 1763), established in Florida in 1565, shows about 12% of the area within the city walls devoted to orchards and 11% for pastures, about half of which is

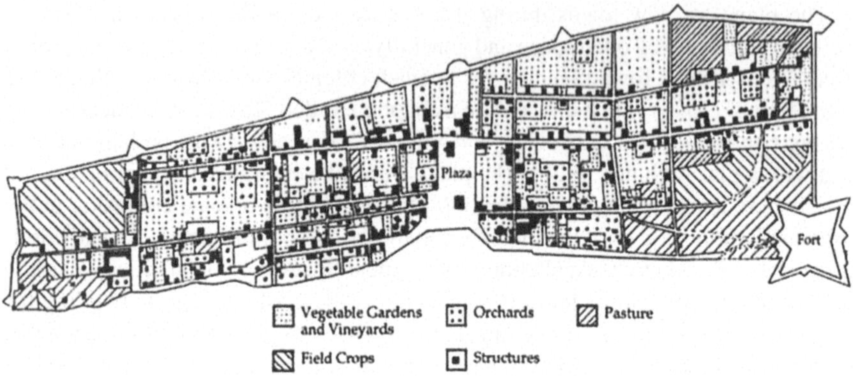

Fig. 4.22 Augustine, Florida (Based on map by Thomas Jeffreys 1764)

planted with scattered trees (Fig. 4.22). These shaded pastures within the walls of the city are reminiscent of the shaded pastures common to peasant closes at the end of the Middle Ages described by Landsberg (1995).

In the British colonies of the new world, towns were initially laid out in either a rectilinear (i.e., Plymouth, Cambridge) or linear (i.e., Salem, Springfield) pattern (Fries 1977). The rectilinear pattern was the result of deliberate planning, while the linear pattern evolved as a result of organic growth. The linear pattern was often in response to the location of streams or to topographic or other natural boundaries in the landscape (Powell 1965). Unlike the Spanish colonists, the early English settler did not have a conceptual plan, provided by British Crown, for their new cities. The basis for town planning by the Puritans grew out of their familiarity with rural villages in East Anglia and Old Testament precedents for the "ideal city of God" (Fries 1977). In 1635, a document entitled "Essay on the Laying Out of Towns", believed to be written by Thomas Graves (the official surveyor of the Massachusetts Bay Company), describes a town plan based on concentric squares (Massachusetts Bay Record 1861). This document influenced the laying out of nucleated villages that prevailed in New England. Central to these villages, which in some cases later grew into towns and cities, was a commons generally known as the village green (Stilgoe 1982). The village green served as a place where the local militia could drill and a place where livestock were kept overnight to be taken in the morning to the common pasture surrounding the village by municipal official known as the cattle reeve (Mumford 1961). In colonial times trees were not planted on the village greens. Lots around the village green were set aside for a meeting house and later a church and school. The area beyond the village green and its surrounding lots was divided into 0.5–1.5 acre lots for houses and gardens. Orchard trees were reported to be common in these garden plots (Josselyn 1663). Beyond the zone set aside for housing were zones designated for common pastures, farm fields, and wood lots. These zones were sometimes laid out in a circular pattern as well a pattern of concentric squares proposed in the "Essay on the Laying Out of Towns". There is no evidence to suggest that street trees were initially planted in these early American villages.

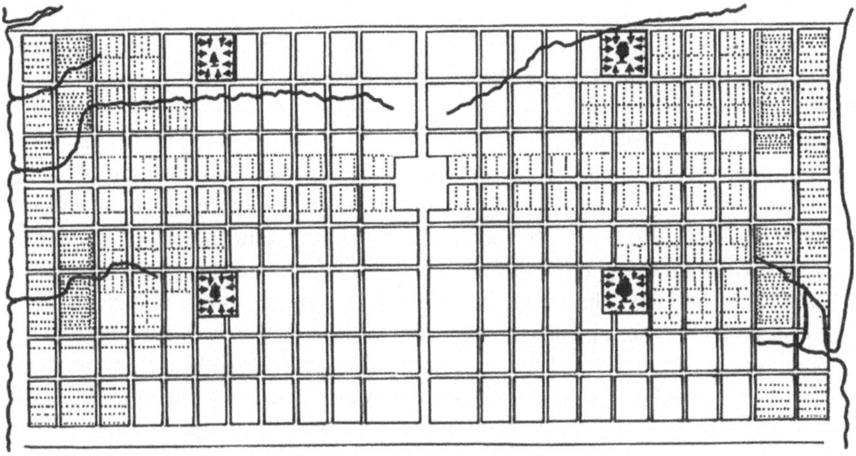

Fig. 4.23 Plan for Philadelphia, 1682 (Redrawn from Thomas Holme (Graven 1963)

The Puritan concept of town planning did not prevail for long. Stilgoe (1982) suggests the nucleated town plan prescribed under the “Essay on the Laying Out of Towns” was being abandoned by the middle of the seventeenth century. Plans for the towns of Wethersfield, Connecticut (1640); Salem, Massachusetts (1670); and Springfield, Massachusetts (1640) illustrate non-nucleated patterns.

Philadelphia was the first town planned by English colonists to set aside areas for parks and to prescribe the planting of trees in these parks. It was planned as the seat of government of William Penn’s land grant in the New World. Thomas Holme drew a plan for the city in 1682, which included five public squares within a grid of residential blocks (Fig. 4.23). The central square was designed for the mustering of the militia while the other four squares were for the enjoyment of people living around them (Garvan 1963). These squares are shown on the plan with trees. Individual lots were one-half acre that was large enough to support a house, garden and small orchard. William Penn had wanted Philadelphia to be a “greene country town” (Penn 1681) and the provision for tree lined squares and space for individual orchards was a way of achieving his objective.

The Enlightenment

The eighteenth century in Europe and the Americas is known as the Enlightenment. It was a century of scientific and intellectual development that fostered a confidence in human reason. A widespread belief in a rational approach to solving the problems of society led many to believe in a general sense of progress and perfectibility. The economic growth of the urban middle class during the eighteenth century led to pressures for political reform culminating in the American and French revolutions.

A rising middle class also sought amenities in urban areas that had previously been available only to the aristocracy. Among these amenities were pleasure gardens where people could stroll in the shade of trees, admire flowering plants and song birds, be entertained by various performers, bowl, dance, eat and drink, and probably of significant importance, see and be seen by their peers. The first pleasure gardens appeared in London sometime before the English Restoration (1660), but they grew in size, prominence, and popularity in the eighteenth century (Wroth 1896). They were initially established on the sites of private gardens or sometimes nurseries where they could take advantage of existing tree lined avenues, allees, and arbors. These provided shaded walks that were among the attractions of the more famous London pleasure gardens. Samuel Pepys described June 7, 1665 in his diary as "the hottest day that ever I felt in my life". To survive that heat he went to the Spring Garden at Foxhall and spent the afternoon and early evening pleasantly walking and drinking the water. The Foxhall garden Samuel Pepys visited in 1665 became the nucleus for the famous Vauxhall gardens of eighteenth century London. Additional plantings were added to the Foxhall garden to create a pleasure garden with two 900 foot long avenues defined by elm trees, extensive orchards of cherry trees, and an area of densely planted trees known as the wilderness (Fig. 4.24) (Wroth 1896).

Sixty-four pleasure gardens were in operation in London during the eighteenth century (Wroth 1896). The largest of these, like the Vauxhall gardens, were several acres in size. Commonly cited trees include common lime, mulberry, holly, willow, elder, yew, poplar, elm, London plane trees and many fruit trees including cherry, pear and apple. These gardens served the recreational functions of the pall mall courts and promenades of the Renaissance. Initially the London pleasure gardens were opened to the public free of charge. The owners of the gardens profited by the

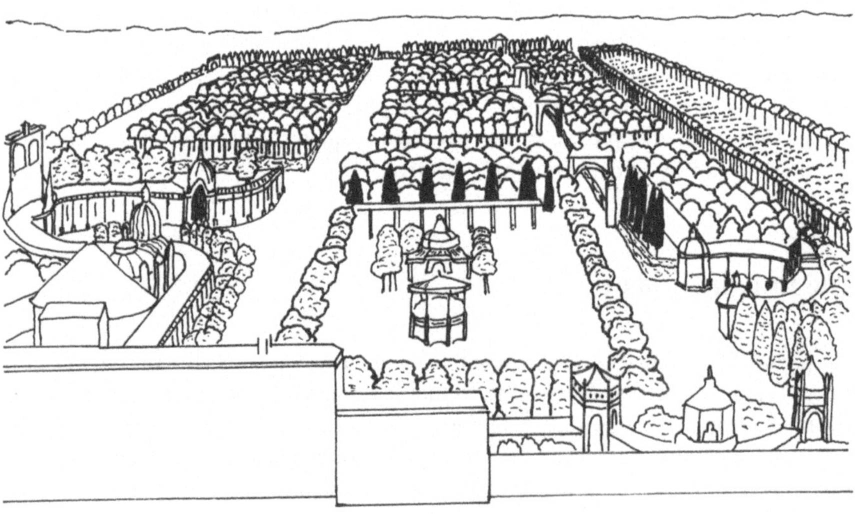

Fig. 4.24 Vauxhall Garden, 1751 (Redrawn from Wroth 1896)

sale of tea, wine, cakes, and other items of food. These could be consumed at tables often set out between the trees along the avenues. In the early part of the eighteenth century admission was charged at most of the London pleasure gardens to both obtain additions profit and to control the class of garden users. Entry fees ranged from one shilling for the day to three shillings for the season. Members of both the middle class and the aristocracy frequented the pleasure gardens. Although the aristocracy had exposure to trees in the park like settings of royal estates and their own private gardens, the pleasure gardens provided the middle class with an opportunity to stroll beneath shaded avenues of lofty trees and see the many species of fruit trees used in the orchards. The tree-dominated public parks that were to emerge in London and other English cities in the nineteenth century had their origins in the pleasure gardens of London during the enlightenment. The demise of the London pleasure gardens came as a result of opposition to the gambling and drinking that took place in these gardens. Echoes of the pleasure gardens that emerged in London in the seventeenth century can be seen today in the beer gardens of Germany cities, the Tivoli Garden in Copenhagen, and the Norbulingka Lam Park in Lhasa.

Another development influencing the planting of trees in cities also occurred in London in the eighteenth century. This was the creation of residential squares that were incorporated into new housing areas for the emerging upper middle class. They provided the natural amenities of the pleasure gardens without the problems associated with gambling and drinking. Lawrence (1993) credits these residential squares as having a powerful influence on the way we view the urban landscape today. Although the residential square evolved in England it was copied in other parts of Europe and in the emerging cities of the Americas. The first residential square build in London was designed by Inigo Jones. Construction of this square and its adjacent buildings began in 1630. It became known as Covent Garden. Like other early residential squares it was simply paved and did not support trees or other plants. Leasehold houses were available for purchase on two sides of the square. Porticoes fronting these houses provided shade and protection from rain for pedestrians. It was an open square and the public found it a popular place for strolling. Merchants set up stalls and converted a portion of the square into a market. These were not intended uses either by the developer nor the leaseholders. Subsequent residential squares built in the eighteenth century were surrounded with fences and were accessible only through locked gates, although the public was permitted to walk along the streets surrounding these private residential squares. Leicester Square, initiated in 1661, was the first residential square to be planted with trees (Lawrence 1993). The popularity of trees, lawns, and shrubbery planted in Cavendish, Berkeley, and Grosvenor squares stimulated the planting of earlier squares that had simply been paved. Lawrence (1993) suggests, on the basis of his review of several engraving of London's residential squares made in the first half of the eighteenth century, that the plantings in these squares usually consisted of plain turf plants. If trees were included they were usually planted in neat rows. Small trees such as laburnum were most often used, but occasionally lindens and hornbeams were planted. These larger trees were pruned to not obstruct views across the residential square. These improvements usually led to the fencing of the planted areas and attempts to restrict the use of these squares to the surrounding leaseholders. In 1726 Parliament was encouraged to pass an act of enclosure by the leaseholders around St. James Square. This act allowed for the construction of a fence around the planted portion of the square and for a tax to be levied providing for improvements and maintenance of the area within the fence. Subsequent acts of enclosure were passed by Parliament to allow for the fencing and exclusion of the public from other residential squares (Report of the Royal Commission on London Squares 1928). Although the acts of enclosure prevented the general public from having entry into the residential squares, these acts did not initially prevent visual access to the squares.

By the 1720s a distinctly English style of landscaping was emerging on rural estates that influenced the design of London's residential squares (Watkin 1982). This influence did not affect the residential squares of London until the latter half of the eighteenth century. The parterres inspired by the Italian Renaissance gardens and initially used in London's residential gardens were replaced by more naturalistic arrangements of trees or by the planting of trees around the edges of the squares. Berkeley Square that had been enclosed and improved with tree-lined walks in 1766 was replanted in 1777 with London plane trees. These were allowed to grow to their full height and are believed to be the oldest London plane trees currently found in London (Fricker 1974). The redesign of Berkeley square, characterized by the increased tree density and the use of larger stature trees is typical of the changes made in other residential squares in the last half of the eighteenth century. Elms, London plane, and common lime were among the trees usually planted.

By the early nineteenth century most residential squares had been redesigned. The initial formalistic plantings, often resembling Renaissance gardens in their shrub-line walks and allees of trees gave way to the more naturalistic English landscape style. The nineteenth century also saw more peripheral planting of trees and shrubs to insure privacy for the lease holders using the gardens. These were not pruned, as the trees had been in the first half of the eighteenth century to allow for views across the squares, but were encouraged to grow thickly to both obscure views across the squares and to screen the interior of the squares from the public view. Residential squares in Kensington, as well as other parts of London, continue to support dense plantings of trees and shrubs around their margins to maintain the privacy of the interior space for the leaseholders. In contrast, Berkeley square (now a public park) is unencumbered by shrubs or the lower branches of trees allowing one to easily see across the square.

The popularity of residential square spread from London to other cities in Europe, most notably Paris, as well as to the English colonies in the new world. Napoleon III's plan for the remodeling of Paris was influenced by his experience as a young man with the residential squares in London. However, he required the residential squares in Paris be open to the public and be public parks rather than private gardens (Pinkney 1958). Oglethorpe's plan for the city of Savannah, Georgia in 1734, drew directly on the residential squares of London and was initially intended as private gardens (Reps 1969). As an urban form the residential square played an important role in the incorporation of trees into urban areas. It can be seen as a precursor for the courthouse square that was developed in the United States in the nineteenth century (Price 1968) and the small wooded urban park that emerged in American cities in the twentieth century.

The Nineteenth Century

In the nineteenth century developments in Europe and the United States solidified the place of trees in urban environments. Although most of the precedents for the incorporation of trees into urban areas predate this century, it was in the nineteenth century that the remodeling of Paris, the establishment of large public parks in England, and the development of wooded suburbs in both England and the United States stimulated the incorporation of trees in public urban spaces and private residential gardens at a scale previously unknown. The nineteenth century was also a period of renewed interest in "plant hunting" that resulted in the introduction to many new tree species to the urban forests of Europe as well as to cities in North America (Mitchell and House 1999).

The tree-lined boulevard had its origins in the tree-lined processional ways of the ancient world, the county roads lined with trees as early as the third century AD in Japan, the allees and avenues of the Italian Renaissance gardens of the fifteenth century, and the malls and promenades of the late sixteenth and early seventeenth centuries European cities (Lawrence 1988). However, transformation of Paris during the reign of Napoleon III, under the direction of Haussmann, codified the tree lined boulevard as an essential requirement for cities in the later half of the nineteenth and twentieth centuries. Much has been written about the motivations for the creation of the boulevard system in Paris (Alphand 1868-73; Pinkney 1958; Kostof et al. 1979; Girouard 1985; Jacobs et al. 2002). The remodeling of Paris may have been stimulated by the ambitions of Napoleon III to demonstrate his power, stimulate economic development, and for military control over the local population, but it also had at its roots a desire to improve the city and the health of the urban population. Key to this improvement was the increased width of the new thoroughfares, provisions for sidewalks, proper paving, and drainage, and the incorporation of trees. (Fig. 4.25) Haussmann's plan envisioned dozens of new boulevards radiating out from Place d'Etoile, Place d'Trocdero, Place d'Bastille, and Place d'Italie. In all 57 miles of wide streets were constructed, 700 miles of sidewalks installed, and the number of trees along the streets in Paris was doubled (Pinkney 1958). The new boulevards and widened streets were planted primarily with plane trees (*Platanus x hispanica).* Horse chestnut (*Aesculus carnea*), big leaf linden (*Tilia platyphyllos)* and ash (*Fraxinus excelsior)* were also used, but to a lesser extent than the plane tree. These four species were no doubt chosen because of their large stature and arching form that provided shade for both the sidewalks and streets. The tree-lined boulevards in Paris were copied in numerous cities through out the world in the later half of the nineteenth and early twentieth centuries. They continued to serve as

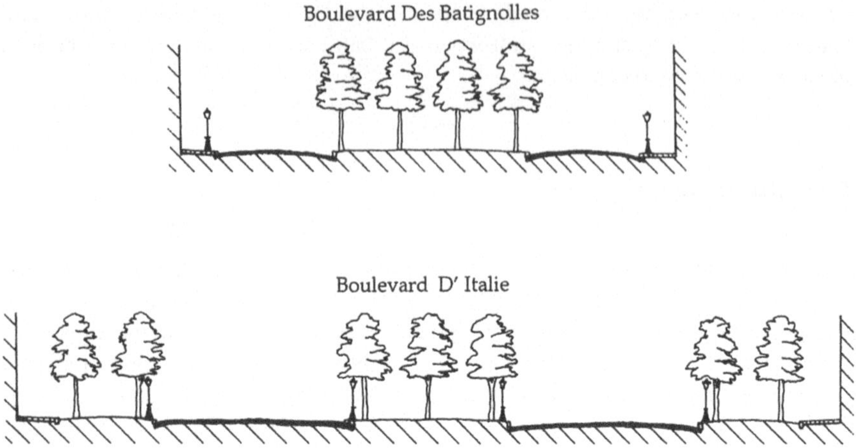

Fig. 4.25 Two boulevards in Paris designed under the direction of Haussmann (Redrawn from Alphand, 1867–1873)

archetype in urban planning in the later half of the twentieth century in new cities such as Brasilia and in the urban expansion of cities in many parts of the world.

Two factors may be credited with the evolution of large public parks during the nineteenth century. The first is the enclosure of residential squares, thus preventing access to the public, and the conviction that large urban parks could improve both the health and behavior of the lower classes. A growing population of people in London, who could not afford to be a leaseholder of houses with access to a residential square, expressed their desire for access to parks in the city. A Parliamentary committee convened to investigate this problem called for the admission of the general public to Regent's Park (Committee on Public Walks 1833). By 1840 Regent's Park was fully opened to the public. It's opening was followed by the creation of Victoria Park in the east end of London to provide additional park space for the public (Chadwick 1966). Similar large urban parks were created in other English cities, notably Birkenhead Park near Liverpool that served as a model for Central Park designed in New York City by Olmsted and Vaux. The London public parks inspired the redesign of the Bois de Boulonge in Paris and the Teirgarten in Berlin (Schuyler 1986). Sexby (1898) has suggested the large naturalistic urban park can be traced back to the royal hunting parks at the edge of London. After the seventeenth century these were increasingly opened to larger segments of the public, repeatedly renovated and redesigned until by the early decades of the nineteenth century they arrived at a form that was imitated by the designers of new urban parks. The form of the urban parks emerging in London in the first half of the nineteenth century also has its roots in the English landscape designs of Capability Brown and Humphrey Repton. Their break with the formal designs of the Renaissance garden and their creation of mosaics of tree groves, meadows, and bodies of water can be seen in the designs of the large public parks of the nineteenth century. The desire to emulate the natural character of the landscape resulted in greater tree species richness in the large public parks of the nineteenth century as compared to the diversity

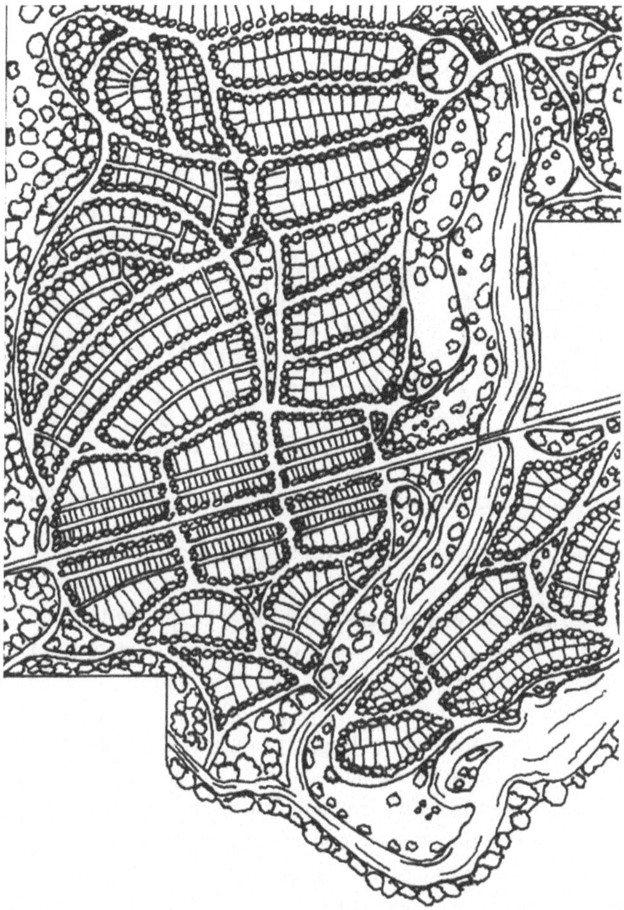

Fig. 4.26 A portion of the general plan of Riverside, Ill (Redrawn from Olmsted and Vaux plan, 1869, in Tishler, 1989)

of trees used in the boscoes of the Italian Renaissance gardens. In 1625 Francis Bacon in his essay "Of Gardens" suggested a list of 21 trees for planting in the Renaissance gardens of northern Europe (Bacon 1625). In contrast Olmsted requested over 150 tree species be planted in Central Park in 1858 (Beveridge and Schuyler 1982).

In the early part of the nineteenth century rural cemeteries were laid out in the United States with tree-lined curving roads. Mount Auburn cemetery on the outskirts of Boston built in 1831 was the first of these new rural cemeteries to break with the traditional grid lay out of earlier American cemeteries (French 1975). Mount Auburn and other rural cemeteries served as an inspiration for suburban development in the second half of the nineteenth century. Olmsted's plan for Riverside, a suburb of Chicago, is often cited as an example of such a suburb inspired by the rural cemetery movement (Fabos et al. 1968). In the Riverside project (Fig. 4.26), Olmsted envisioned curving tree-lined streets as well as open space

Fig. 4.27 Deerfield, Massachusetts, c. 1840 (Redrawn from an engraving by J. W. Barber)

areas adjacent to a creek where trees would also be planted (Rybczynski 1999). Each residential lot also provided space for the home owners to plant additional trees. Suburbs developed around London toward the end of the nineteenth century as rail transport provided a means of moving workers in and out of the city. Semi-detached houses were typical of the English suburbs. The appeal, in part, of these suburbs was the opportunity for home owner to have space for a garden. As in the American suburbs, street trees were planted in these new suburbs. Schein (1993) makes the argument that the street tree is an American invention based of the paucity of street trees in older towns and villages in Europe. The common planting of street trees in American towns and villages after the turn of the nineteenth century (Jackson 1970) may have proceeded the planting of street trees in European suburb that developed at the end of the nineteenth century (Fig. 4.27); however, ample evidence has been presented by Lawrence (1988) for a much earlier origins of tree-lined streets in Europe. The significance of the American and European suburbs as a model for the incorporation of trees into urban areas is not only in the use of tree-lined streets, but in the use of privately owned garden space for trees. American suburbs average 35–50% tree canopy cover depending upon the region in which they are located, much of this cover provided by trees in private space. In older suburbs in the southeastern United States tree canopy cover exceeds 60% (http://www.americanforest.org/resources/urbanforests/treedeficit.php).

The Twentieth Century

The "City Beautiful Movement" of the early twentieth century encouraged the planting of parkways and the establishment of new parks in many cities in the United States (Wilson 1989). The Park and Parkway Plan for Kansas City, Missouri

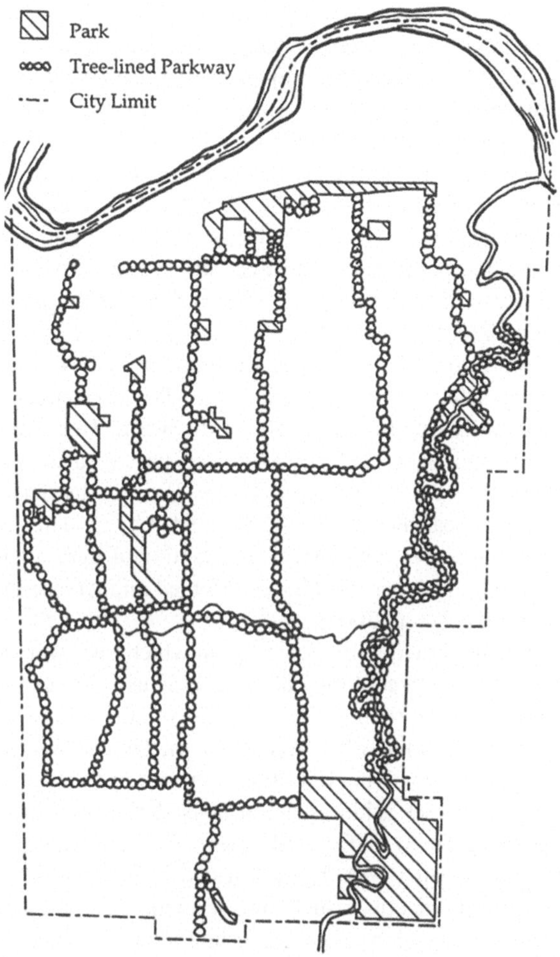

Fig. 4.28 Kansas City Parkway plan, 1915 (Redrawn from Kessler 1915)

(Fig. 4.28) and the Parks and Boulevard System for Seattle, Washington are exceptional examples. The City Beautiful Movement did not provide any particular innovation in terms of the use of trees in cities, but it did focus considerable attention on the value of trees in improving urban environments.

In the later part of the twentieth century the environmental movement, which had initially been concerned with the management of forests and other wildland areas, encompassed urban environments. Environmentalists encouraged the improvement of urban environments through the restoration of parks, increased tree planting, the use of native tree species, and the establishment of heritage tree programs. Various groups emerged, such as the Tree People in Los Angeles and the Friends of the Urban Forest in San Francisco, to develop community-based tree planting programs (http://www.treepeople.org). The maintenance and management of a number of city parks was also passed on to concerned citizen groups who could muster the volunteers and develop financing for restoration projects. Notable among these organizations has been the Central Park Conservancy (http://www.centralparknyc.org).

In spite of the efforts of the City Beautiful Movement and the Environment movement to improve the American urban areas through tree planting, tree density and canopy cover declined during the last half of the twentieth century. Part of this declined can be attributed to street widening projects and the installation and management of utility lines (Schein 1993). Urban forest in the United States were also devastated by the Dutch elm disease in the twentieth century. This epidemic reduced street tree populations and tree canopy cover in many cities in northeastern and Midwestern portions of the United States (Campanella 2003). Insufficient urban budgets for the management of ageing street tree populations have also resulted in a failure to replant all trees that must be removed for reasons of tree health and human safety (Beatty and Hickman 1981). The situation in the United States is not necessarily true for other parts of the world. Tree planting has taken place in many Asian cities as urban renewal projects have created new green space. Conversion of extensive areas of one and two story housing districts in Shanghai and Beijing into high rise apartment building surrounded by gardens has increased tree density and canopy cover in both cities. The planting of trees along many streets that formerly were without trees has also contributed an increase in tree cover in many Chinese cities (Yang et al. 2005).

It is hard to identify new ways in which trees were incorporated into urban areas in the twentieth century. The street tree, use of trees in parks and private gardens as models of tree use had all evolved before the start of the twentieth century. It might be argued that the urban greenbelts, the Dutch "Woonerf", and bioswales are urban innovations involving trees that came out of the twentieth century. Areas of natural forest were set aside as green belts during the growth of cities like Stockholm (Rydburg 1998) or forested greenbelts were planted at the margins of cities like Beijing (Chen 1998). These forested greenbelts have been established to meet an number of objectives, including aesthetics, air pollution reduction, wildlife habitat, recreation, erosion control, and limitation of urban growth (Smith 1993). An outstanding example of the employment of forested greenbelts for the improvement of air quality can be seen in the system of greenbelts developed by the city of Stuttgard, Germany. These greenbelts utilized conifer trees to clean the natural up-canyon afternoon winds flowing into the city from industrial areas in the Neckar Valley. Greenbelts at elevations above the city functioned to cleanse air moving down the Valley at night (Hoffman 1976).

Trees were used use as traffic barriers in the shared street or "woonerf" designed by De Boer in the Netherlands (Southworth and Ben-Joseph 1997). The design of a woonerf utilized trees so that the motorist would feel as if the driver were driving in a "garden" setting, thus requiring the recognition of other users of the street (Fig. 4.29). The woonerf allowed the owners of automobiles to drive to their houses, but also converted the street into a space that could be more safely used by children for play (Jonquiere 1978). Trees were essential to the design of the woonerf because they helped to create the garden atmosphere and they served as physical barrier to slow the traffic. Trees served a similar purpose in the traffic "calming" circular

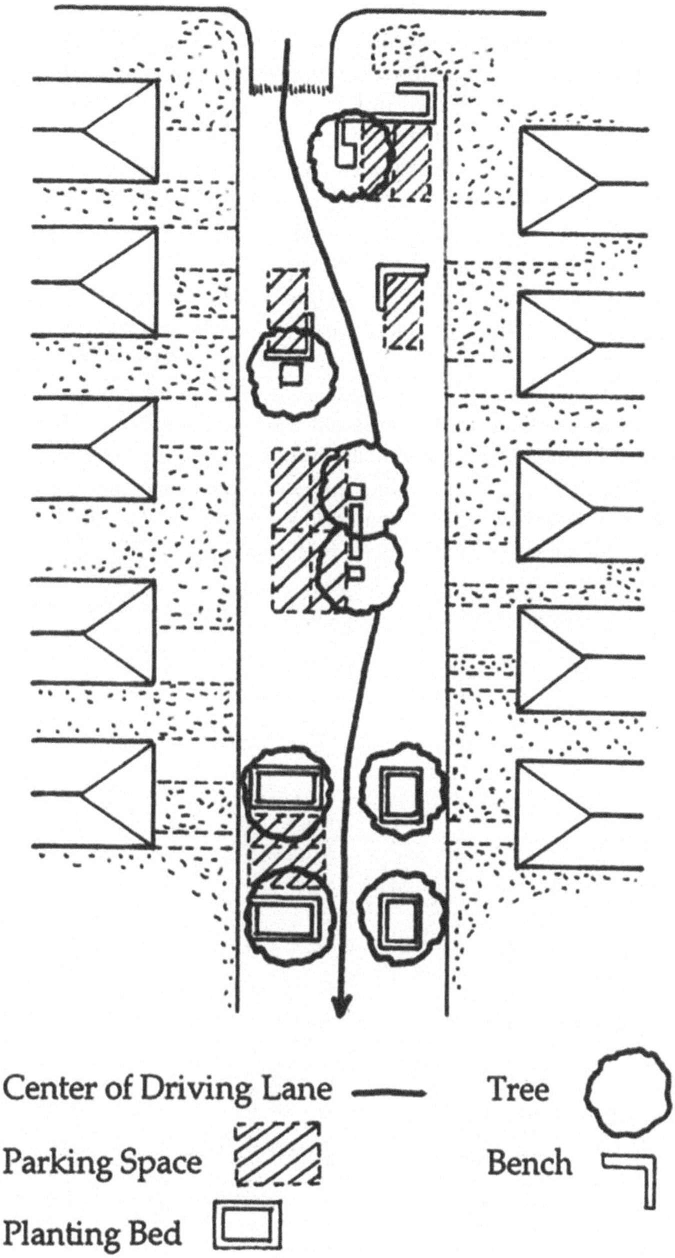

Fig. 4.29 Woonerf (Redrawn from Southworth and Ben-Joseph 1997)

islands installed at street intersections in Seattle and other cities. In Seattle, three trees were usually planted in these traffic islands to provide both a physical barrier and a visual reminder that the motorist was approaching an intersection.

Bioswales were first constructed in Portland, Oregon in the late 1990s. A bioswale is a drainage course designed to remove silt and pollution from surface runoff water (France 2002). Trees such as tupelo (*Nyssa* sylvatica) and other species that are tolerant of periodic flooding are commonly in the bioswales to reduce the velocity of water. The presence of trees in bioswales also adds to the aesthetics of the street along which bioswales are installed (Fig. 4.29).

In thinking about the history of the use of trees in urban areas one can be reminded of a line from William Faulkner book *Absalom, Absalom!* (1936) "The past is never dead, it's not even past". The ways in which trees have been used in urban areas evolved with the evolution of the city and these uses continue to be employed today. These "past" uses continue in their original form and sometimes show modifications with time. In spite of these modifications one can usually recognize the historical precedent for the use. Understanding this evolution adds to our appreciation and understanding of the urban forest. It may also serve to stimulate future uses of trees in urban areas that inevitably will be well rooted in the past.

Chapter 5
Composition

Surveys of trees in the 33 cities studied provide information for contrasting tree species composition in different urban environments. The cities surveyed are representative of large cities in each of the biomes. Although they may not encompass all of the urban forest variations within each of the biomes, the conclusions drawn from these cities are expected to illustrate the general composition of the world's urban forest. The information collected has been used to make some conclusions about species diversity and frequency in different urban environments. The relationship of species diversity and frequency to the ecology of the biomes and the influence of cultural and historical factors will be examined in a subsequent chapter.

The tree species that are present in urban forests are the result of the conscious introduction of trees, decisions to maintain trees from the pre-existing urban vegetation, and natural establishment of trees within cities (McBride and Jacobs 1986). The ability of a tree species to survive in a city is determined by the climate, although maintenance techniques such as irrigation may allow certain species to survive in climates where they do not naturally occur. Trees occurring along streets and boulevards have for the most part been selected and planted by agencies and individuals responsible for streets and boulevards. Parks in forested biomes occasionally support individual trees and groves that are relicts of the natural vegetation. In some cases trees found in parks are the offsprings of native trees that previously occupied the site. Parks may also occur on sites where no trees previously existed and the trees one observes in these parks are the result of conscious species selection, planting, and maintenance. Waste areas, untended gardens, cemeteries, and transportation corridors are environments in one may find a combination of trees previously planted in these environments and trees that were naturally established.

The number of tree species found in most cities varies from a few dozen to over 1000 depending on the climate and the history of urban design. In Nuuk, within the Arctic circle, climate limits the urban forest to only five tree species. In contrast, Rio de Janeiro has an urban forest composed of more than 2000 species due to its tropical climate, the occurrence of trees native to the Atlantic rain forest found in the natural parks within the city, and the influence of the Brazilian landscape architect

J.R. McBride, *The World's Urban Forests*, Future City 8,
DOI 10.1007/978-3-319-52108-4_5

Roberto Burle Marx. Marx was dedicated to the introduction of native Brazilian species to Rio de Janeiro and other parts of the world (Montero 2001). His plan for Parque do Flamengo, built in 1965, called for the planting of 300 tree species. These were not planted as individual specimens, as one might see in a botanical garden, but in groves and along thoroughfares and walkways. Parque do Flamengo now supports over 170,000 trees.

Overall Composition of the Urban Forest

In the 33 cities surveyed for this book a total of 450 tree species were recorded. The greatest number was recorded in Vancouver (64 species) and the least in Nuuk (5 species) (Table 5.1). The number of species recorded in the surveys of each of the cities is an under-estimate of the total number of species present. No sampling procedure, short of a 100% inventory, could identify all of the tree species in a city. This study was not intended to estimate the total number of tree species, but to identify the more common species used in different public urban environments. In eight cities, where records were available, the number of species recorded in the surveys averaged 39% of the number of species reported to be in these cities (Table 5.2). In cities, such as Murmansk, where to number of tree species occurring was low (16 species) a higher percentage of tree species were observed in the survey (81%). Cities in which a large number of trees species were used the survey recorded a smaller percentage of species. For example, Beijing with 170 tree species of which only 19% were recorded in the survey.

Although most cities have large numbers of tree species, relatively few species are commonly used. For example, in Chicago five species make up 75.7% of the tree population (Nowak 1994a, b), in Beijing the five most common species constituted 60.8% of the tree population (Yang et al. 2005), and in Hong Kong the five most common species represent 47% of the street tree population (Jim 1999). In a survey of Los Angeles only 28 of a total of 175 species that each comprised more than 1% of the total tree population (McPherson et al. 2003; Hodel 2005). In many cities a very few tree species have been used as street trees. The species used along streets often dominated the percentage of trees in an urban forest because the population of street trees out number the population of trees in parks and other urban environments.

An attempt was not made to determine the total number of trees in the cities in this study. The large sample size necessary to establish the species composition of an urban forest was beyond the scope of this book. Nowak et al. (2003) has suggested a minimum sample of 200 plots is needed to determine percent species composition in large cities. Time and resources were not available to sample at that intensity. Only 30–50 sample sites were surveyed in each city for this study.

The results of the survey of trees in different urban environments indicate that a majority of the species used are limited to individual cities. A total of 450 species were identified in the survey. Two hundred and fifty-one of these species (55.7%)

Table 5.1 Number of tree species recorded

Biome	City	Number of species
Tropical forest	Bangkok	36
	Rio de Janeiro	50
	Singapore	39
Broadleaf evergreen forest	Hong Kong	61
	Sydney	35
	Tokyo	33
Deciduous forest	London	46
	New York	59
	Paris	39
Coniferous forest	Jacksonville	41
	Seattle	45
	Vancouver	64
Mixed deciduous-coniferous forest	Beijing	35
	Moscow	40
	Stockholm	29
Mediterranean woodland/scrub	Athens	34
	Los Angeles	79
	Tel Aviv	48
Savanna/scrub	Brasilia	42
	New Delhi	54
	Lagos	25
Grassland	Buenos Aires	39
	Johannesburg	53
	Tehran	31
Desert	Cairo	38
	Dubai	28
	Phoenix	44
Tundra	Murmansk	13
	Nuuk	5
	Reykjavik	12
Highland	Cusco	38
	Lhasa	18
	Quito	49

were observed only one city. These species of limited urban occurrence are analogous to naturally occurring endemic species restricted to limited geographical areas by soil, micro climates, or barrier to migration. Many of the 251 species of limited urban distribution were often native to the region of the city in which they occurred. The ghaf tree *(Prosopis chineraria)* observed in Dubai, the Florida maple *(Acer barbatum)* in Jacksonville, or the Tibetan juniper *(Juniperus tibetica)* in Lhasa are examples of trees native to the local biomes in which these cities occur. These local native species have not been introduced into the urban forests of other cities. The

Table 5.2 Relationship between the number of tree species reported and observed in eight cities

Biome	City	Number of tree species Reported	Observed	Percent of reported
Tropical forest	Bangkok	127[1]	33	26
Broadleaf evergreen forest	Hong Kong	149[2]	60	40
Deciduous forest	New York	168[3]	61	36
Mixed conifer-deciduous forest	Beijing	170[4]	32	19
Grassland	Buenos Aires	160[5]	40	25
Desert	Dubai	85[6]	30	35
Tundra	Murmansk	16[7]	13	81
Highland	Quito[a]	25[8]	13	52

[a]Trees in boulevard medians only
[1]Thaiutsa 2000; [2]Jim 1996; [3]Pepper et al. 2007; [4]Beijing Annals Editor Board 2000; [5]Valla et al. 1999; [6]Sharif; [7]Kazakov; [8]Gangotena et al. 1990

remaining 199 species occur in at least two of the 33 cities. Ten of these species were observed ten or more cities (Table 5.3). These widely used species are characterized by a tolerance for environmental conditions existing in different biomes, a particular fitness for survival in the urban environment, the ability to be useful in specific urban environments, and visually attractive characteristics. All ten of the most commonly encountered species occurred in cities located in at least five of the eleven biomes. Weeping willow *(Salix babylonica)* was recorded in cities in nine of the eleven biomes and 16 of the 33 cities visited in the study. The widespread use of weeping willow results from its unique form as well as its adaptation to environmental conditions in several biomes. Southern magnolia (*Magnolia grandiflora*) was also observed in 16 of the 33 cities, but only eight of the eleven biomes. Southern magnolia (*Magnolia grandiflora)*, jacaranda *(Jacaranda mimosifolia)* and royal poinciana *(Delonix regia)* are valued for their beautiful flowers and are planted in and along streets in many parts of the world. London plane *(Platanus acerifolia)*, black locust (*Robinia pseudoacacia)*, and white mulberry (*Morus alba)* are widely used as street trees and are noted for their tolerance to urban environments. Italian cypress (*Cupressus sempervirens)*, royal palm *(Oreodoxa regia)*, and Mexican fan palm *(Washington robusta)* are listed among the ten most commonly recorded species in large part because of use in specific urban environments. Italian cypress is frequently planted in cemeteries throughout the temperate zone for its somber-evergreen foliage. Royal palm and Washington palm have found an important niche as trees for planting in boulevard medians. Their roots can accommodate the narrow street median and their crown canopies are above the ongoing traffic.

One way of characterizing the tree species composition of a city, when actual tree population's numbers are not available, is to calculate the percentage of sample sites at which a given species was recorded. This measures the dispersion of tree species across the different environments in a city and is called the frequency of occurrence. It may also be calculated for specific environments, such as streets or parks, to determine the relative importance of a species in different urban environments.

Table 5.3 Tree species recorded in ten or more cities

Biome	City	Black Locust	Southern Magnolia	Weeping Willow	Jacaranda	Italian Cypress	Royal Poinciana	Royal Palm	White Mulberry	Washington Palm	London Plane
Tropical forest	Bangkok			•				•			
	Rio de Janeiro				•		•	•			
	Singapore				•		•	•			
Broadleaf evergreen forest	Hong Kong			•	•		•	•	•	•	•
	Sydney	•			•	•		•			•
	Tokyo	•		•					•		•
Deciduous forest	London	•		•		•					•
	New York	•	•						•		•
	Paris	•			•	•	•			•	•
Coniferous forest	Jacksonville		•	•						•	
	Seattle	•	•	•							
	Vancouver	•	•	•						•	•
Mixed deciduous-coniferous forest	Beijing	•		•					•		
	Moscow	•									
	Stockholm		•	•							
Mediterranean woodland/scrub	Athens	•	•		•	•	•		•	•	•
	Los Angeles	•	•		•	•			•	•	•
	Tel Aviv		•		•	•	•	•	•		
Savanna/scrub	Brasilia		•		•		•	•			
	Lagos					•	•	•		•	
	New Delhi		•	•			•	•	•		
Grassland	Buenos Aires	•	•	•	•	•	•	•		•	•
	Johannesburg	•	•		•	•	•	•	•	•	•
	Tehran	•	•	•		•			•		

(continued)

Table 5.3 (continued)

Biome	City	Black Locust	Southern Magnolia	Weeping Willow	Jacaranda	Italian Cypress	Royal Poinciana	Royal Palm	White Mulberry	Washington Palm	London Plane
Desert	Cairo		•	•	•	•	•	•	•	•	
	Dubai						•			•	
	Phoenix				•	•	•	•	•	•	
Tundra	No tree species occurring in Murmansk, Nuuk, or Reykjavik occurred in ten or more cities										
Highland	Cusco		•	•		•					
	Lhasa			•							
	Quito		•	•	•	•				•	•
# Biomes		6	8	9	8	7	7	6	7	6	6
# Cities		14	16	16	14	14	14	13	12	11	12

• means the species is present

The five tree species that occurred most frequently in each city are shown in Table 5.4 Only the top five species are listed because the frequencies drop off sharply after the first five. Even among the top five species one can see low frequencies of occurrence, such as the 6% frequency of royal palm *(Oreodoxa regia)* in Singapore and the 9% frequency of coconut palm *(Cocos nucifera)* in Rio de Janeiro. The five species with the highest frequencies of occurrence would be expected to include the most common species in a city based of tree population studies. For example, Thaiutsa et al. (2000) reports the five most common species in Bangkok (on the basis of tree population) are angsana (*Pterocarpus indicus),* tamarind *(Tabebuia rosea),* golden shower *(Cassia fistula),* Honduras mahogany *(Swietenia macrophylla),* and bullet wood *(Mimusops elengi).* All of these were recorded in samples taken in Bangkok and three were among the top five in terms of frequency of occurrence. In Beijing the five most common species are Japanese pagoda tree (*Sophora japonica),* Chinese white poplar *(Populus tomentosa),* Chinese juniper *(Juniperus chinensis),* black locust (*Robinia pseudoacacia),* and Chinese red pine (*Pinus tabulaeformis)* (Yang et al. 2005). The survey of Beijing conducted for this book found four of these five species among the top five species on the basis of their frequencies of occurrence.

The frequencies of occurrence of the top five species ranged from a high of 91% for silver birch *(Betula pendula)* and rowan *(Sorbus aucuparia)* in Murmansk to a low of 6% for royal palm in Singapore. The frequency of occurrence of a species depends on the total number of species used in the city's urban forest and the consistency of the use of an individual species. In biomes where the climate limits the number of tree species one expects to observe higher frequencies. Likewise in cities where there is high preference for a favorite species, the frequency of occurrence of that species can be expected to be high in contrast to less favored species. High frequencies of occurrence will also be recorded in cities where trees species are uniformly distributed over the city.

The lowest frequency of occurrence for the any of the most common species was 16% recorded for London plane tree in Sydney. The urban forest of Sydney is quite diverse and is characterized by the use of London plane along major boulevards and avenues, while most residential streets, parks, cemeteries, and other environments have been planted with other species. In contrast, the very high frequencies of occurrence of silver birch and rowan in Murmansk (both 91%) results from the limited number of trees that can be grown in Murmansk and the uniform planting of these two species throughout the city.

Eight of the cities have individual tree species with frequencies greater than 66% indicating a widespread use of these species in the particular cities (Table 5.5). The extensive use of a single species in a city may result from a local preference for a species native to the region, preference for species that has had a track record for dependability and service, and/or climatic restriction placed on the number of species that can be used in a city. The widespread use of live oak (*Quercus virginiana)* in Jacksonville and oriental plane tree *(Platanus orientalis)* in Tehran are examples of local preference for native species. In Jacksonville live oak has a frequency of occurrence of 80%, while the frequency of occurrence of oriental plane tree in

Table 5.4 The five most common tree species in each city studied and the frequency of their occurrence

Biome	City	Top five species		
		Common name	Scientific name	Freq.*
Tropical forest	Bangkok	Tamarind	*Tamarindus indica*	34
		Royal Poinciana	*Delonix regia*	20
		Pink Tecoma	*Tabebuia rosea*	17
		Honduras mahagony	*Swietenia macrophylla*	14
		Angsana	*Pterocarpus indicus*	11
	Rio de Janerio	Gold medallion tree	*Cassia leptophylla*	38
		Royal Palm	*Oreodoxa regia*	25
		Floss silk tree	*Chorisia speciosa*	16
		Morton Bay Fig	*Ficus macrophylla*	10
		Coconut Palm	*Cocos nucifera*	9
	Singapore	Angsana	*Pterocarpus indicus*	32
		Yellow Flame	*Tipuana tipu*	29
		Rubber Tree	*Ficus elastica*	13
		Coconut Palm	*Cocos nucifera*	10
		Royal Palm	*Oreodoxa regia*	6
Broadleaf evergreen forest	Sydney	London Plane Tree	*Platanus* x *hispanica*	16
		Prickley-leaved Tea	*Melaleuca styphelioides*	14
		Moreton Bay Fig	*Ficus macrophylla*	14
		Swamp Mahogany	*Eucalyptus robusta*	11
		Royal Palm	*Oreodoxa regia*	10
	Tokyo	Ginkgo	*Ginkgo biloba*	39
		Trident Maple	*Acer buergerianum*	23
		Oriental Plane Tree	*Platanus orientalis*	15
		Japanese Maple	*Acer palmatum*	15
		Zelkova	*Zelkova serrata*	15
	Hong Kong	Indian Laurel Fig	*Ficus retusa*	20
		Chinese Banyan	*Ficus benjamina*	14
		Kapok Tree	*Bombax ceiba*	14
		Chinese Juniper	*Juniperus chinensis*	11
		Candlenut Tree	*Aleurites moluccana*	9

(continued)

Table 5.4 (continued)

Biome	City	Top five species		
		Common name	Scientific name	Freq.*
Deciduous forest	London	London Plane Tree	*Platanus* x *hispanica*	51
		Common Lime	*Tilia europea*	20
		Honey Locust	*Gleditsis triacanthos*	13
		English Holly	*Ilex aquifolium*	13
		English Elm	*Ulmus campestris*	10
	New York	Honey Locust	*Gleditsis triacanthos*	29
		London Plane Tree	*Platanus* x *hispanica*	26
		Norway Maple	*Acer platanoides*	26
		Basswood	*Tilia americana*	24
		Pin Oak	*Quercus palustris*	18
	Paris	London Plane Tree	*Platanus acerifolia*	56
		Red Horse Chestnut	*Aesculus carnea*	37
		Honey Locust	*Gleditsis triacanthos*	26
		Norway Maple	*Acer platanoides*	18
		Common Lime	*Tilia europea*	18
Coniferous forest	Jacksonville	Live Oak	*Quercus virginiana*	80
		Cabbage Palmetto	*Sabal palmetto*	54
		Laurel Oak	*Quercus laurifolia*	44
		Southern Magnolia	*Magnolia grandiflora*	32
		Crape-Myrtle	*Lagerstroemia indica*	27
	Seattle	Norway Maple	*Acer platanoides*	19
		Sugar Maple	*Acer saccharun*	16
		Silver Birch	*Betula pendula*	16
		Basswood	*Tilia americana*	16
		Washington Hawthorn	*Crataegus cordata*	16
	Vancouver	Western Red Cedar	*Thuja plicata*	30
		Japanese Flowering Cherry	*Prunus serrulata*	26
		Norway Maple	*Acer platanoides*	18
		European Beech	*Fagus sylvatica*	18
		Douglas-fir	*Pseudotsuga menziesii*	16

(continued)

Table 5.4 (continued)

Biome	City	Top five species		
		Common name	Scientific name	Freq.*
Mixed deciduous-coniferous forest	Beijing	Japanese Pagoda Tree	*Sophora japonica*	50
		Chinese White Poplar	*Populus tomentosa*	33
		Chinese Juniper	*Juniperus chinensis*	29
		Weeping Willow	*Salix babylonica*	21
		Black Locust	*Robinia pseudoacacia*	17
	Moscow	Common Lime	*Tilia europea*	59
		Box Elder	*Robinia pseudoacacia*	39
		European Ash	*Fraxinus excelsior*	37
		Carolina Poplar	*Populus canadensis*	33
		Norway Maple	*Acer platanoides*	26
	Stockholm	Common Lime	*Tilia europea*	46
		Norway Maple	*Acer platanoides*	31
		Dutch Elm	*Ulmus hollandica*	27
		White Birch	*Betula pendula*	27
		Rowan	*Sorbus aucuparia*	19
Mediterranean woodland/scrub	Athens	Canary Island Date Palm	*Phoenix canariensis*	32
		White Mulberry	*Morus alba*	27
		Victorian-Box	*Pittosporum undulatum*	27
		Italian Cypress	*Cupressus sempervirens*	19
		Carolina Poplar	*Populus canadensis*	18
	Los Angeles	Mexican Fan Palm	*Washingtonia robusta*	28
		Bailey Acacia	*Acacia baileyana*	19
		Southern Magnolia	*Magnolia grandiflora*	19
		Indian Laurel Fig	*Ficus retusa*	19
		Canary Island Date Palm	*Pinus canariensis*	17
	Tel Aviv	Indian Laurel Fig	*Ficus retusa*	38
		Mexican Fan Palm	*Washingtonia robusta*	27
		Athel	*Tamarix aphylla*	20
		Cassia	*Cassia fistula*	13
		Canary Island Date Palm	*Phoenix canariensis*	11

(continued)

Table 5.4 (continued)

Biome	City	Top five species Common name	Scientific name	Freq.*
Savanna/scrub	Brasilia	Tipu Tree	*Tipuana tipu*	24
		Golden Trumpet Tree	*Tabebuia chrysotricha*	20
		Floss Silk Tree	*Chorisia speciosa*	16
		Crown of Gold Tree	*Cassia excelsa*	16
		Ceibo	*Ceibo erythrina*	16
	Lagos	Umbrella tree	*Schefflera actinophylla*	42
		River Sheoak	*Casuarina cunninghamiana*	18
		Coconut Palm	*Cocos nucifera*	15
		Canary Island Date Palm	*Phoenix canariensis*	15
		Mast Tree	*Polyalthia longifolia* var. *pendula*	12
	New Delhi	Babul	*Acacia nilotica ssp. indica*	18
		Neem	*Azadirachta indica*	14
		Pipal	*Ficus religiosa*	14
		Ashoka	*Saraca indica*	14
		Toon	*Toona sinensis*	11
Grassland	Buenos Aires	Green Ash	*Fraxinus pennsylvanica*	67
		London Plane Tree	*Platanus x hispanica*	61
		Paraiso	*Melia azedarack*	58
		Tipu Tree	*Tipuana tipu*	36
		Jacaranda	*Jacaranda mimosifolia*	30
	Johannesburg	London Plane Tree	*Platanus* x *hispanica*	36
		Jacaranda	*Jacaranda mimosifolia*	27
		Chinese Elm	*Ulmus parvifolia*	18
		English Oak	*Quercus robur*	16
		Canary Island Date Palm	*Phoenix canariensis*	13
	Tehran	Oriental Plane Tree	*Platanus orientalis*	83
		Eldarica Pine	*Pinus eldarica*	50
		Persian Ash	*Fraxinus oxycarpa*	33
		Arizona Cypress	*Cupressus arizonica*	33
		European Redbud	*Cercis siliquastrum*	22

(continued)

Table 5.4 (continued)

Biome	City	Top five species		
		Common name	Scientific name	Freq.*
Desert	Cairo	Indian Laurel Fig	*Ficus microcarpa nitida*	71
		Royal Poinciana	*Delonix regia*	63
		Rubber Tree	*Ficus elastica*	29
		Royal Palm	*Oreodoxa regia*	18
		Palmyra Palm	*Borassus flabellifer*	13
	Dubai	Canary Island Palm	*Phoenix canariensis*	61
		Jujuba	*Ziziphus spina-Christi*	19
		Mexican Fan Palm	*Washingtonia robusta*	16
		Royal Poinciana	*Delonix regia*	16
		Lebbek	*Albizia lebbeck*	16
	Phoenix	Mexican Fan Palm	*Washingtonia robusta*	42
		Screw-bean Mesquite	*Prosopis pubescens*	37
		Italian Stone Pine	*Pinus pinea*	32
		Jacaranda	*Jacaranda mimosifolia*	29
		Arizona Ash	*Fraxinus velutina*	26
Tundra	Murmansk	Silver Birch	*Betula pendula*	91
		Rowan	*Sorbus aucuparia*	91
		Sallow	*Salix caprea*	45
		European Aspen	*Populus tremula*	36
		Siberian Larch	*Larix decidua* ssp.*siberia*	18
	Nuuk	White Willow	*Salix glauca*	88
		Norway Spruce	*Picea abies*	12
		Siberian Larch	*Larix laricina*	6
		Lodgepole Pine	*Pinus contorta*	6
		Gray Alder	*Alnus incana*	6
	Reykjavik	Rowan	*Sorbus aucuparia*	71
		Black Cottonwood	*Populus trichocarpa*	43
		Silver Birch	*Betula pendula*	43
		European larch	*Larix decidus*	28
		Whitebeam	*Sorbus intermedia*	28

(continued)

Table 5.4 (continued)

Biome	City	Top five species Common name	Scientific name	Freq.*
Highland	Cusco	Monterey Pine	*Pinus radiata*	24
		Blue Gum	*Eucalyptus globulus*	19
		Common Coral Tree	*Erythrina cristagalli*	16
		Carolina Poplar	*Populus x canadensis*	16
		Chinese Toon	*Cedrela sinensis*	16
	Lhasa	Chinese White Poplar	*Populus tomentosa*	67
		Taiwan Pine	*Pinus tabuliformis*	21
		Large Leaf Willow	*Salix daphnoides*	21
		Weeping Willow	*Salix babylonica*	21
		Chinese Juniper	*Juniperus chinensis*	17
	Quito	Blackwood acacia	*Acacia melanoxylon*	33
		Black Elderberry	*Sambucus nigra*	27
		Blue gum	*Eucalyptus globules*	26
		Dealbata acacia	*Acacia dealbata*	18
		Carolina poplar	*Populus x canadensis*	18

Table 5.5 Cities with trees frequencies over 66%

Biome	City	Species Scientific name	Common name	Frequency
Conifer forest	Jacksonville	*Quercus virginiana*	Live Oak	80
Grassland	Buenos Aires	*Fraxinus pennsylvanica*	Green Ash	67
	Tehran	*Fraxinus pennsylvanica*	Oriental Plane Tree	83
Desert	Cairo	*Ficus microcarpa nitida*	Indian Laurel Fig	71
Tundra	Reykjavik	*Sorbus aucuparia*	Rowan	71
	Nuuk	*Salix glauca*	White Willow	88
	Murmansk	*Betula pendula*	Silver Birch	91
		Sorbus aucuparia	Rowan	91
Highland	Lhasa	*Populus tomentosa*	Chinese White Poplar	67

Tehran was 83%. These native species exhibit adaptation to their local environments as well as the growing conditions common to urban areas. The urban use of each species is characterized by ease of planting, rapid rates of growth, and high value for shade. They were both favorite trees for planting by the early residents of both cities. Live oak is widely used in the American South for a shade tree, while the value of oriental plane tree as a shade tree was acknowledged by Marco Polo in the

Table 5.6 Tree species, among the five species with the highest frequencies of occurrence in a city, that occur in four or more cities

Tree species		
Common name	Scientific name	Number of cities
London Plane	*Platanus* x *hispanica*	6
Norway Maple	*Acer platanoides*	6
Canary Island Date Palm	*Phoenix canariensis*	6
Common Lime	*Tilia europea*	4
Royal Palm	*Oreodoxa regia*	4
White Birch	*Betula pendula*	4
Carolina Poplar	*Populus* x *canadensis*	4

thirteenth century when he visited towns in the vicinity of what is now Tehran (Yule and Cordier 1992).

The widespread use of Chinese white poplar (*Populus tomentosa*) in Lhasa, where it has a frequency of occurrence of 67%, is the result of its recent planting by Chinese government as part of a program to modernize the city. Chinese white poplar is widely used in Beijing and other Chinese cities due to its ease of planting, rapid growth, and tolerance of urban environments. It grows readily in the climate and soils of Lhasa and is rapidly becoming a signature tree of the city.

An examination of the five species in each city with the highest frequencies of occurrence reveals the common use of 7 of these species among some of the 33 cities studied (Table 5.6). The three most prevalent species were London plane tree (*Platanus* x *hispanica*), Norway maple (*Acer platanoides*) and Canary Island date palm (*Phoenix canariensis*). Each occurred among the top five species in six of the 33 cities studied. London plane tree has the highest frequency of occurrence in four of the six cities in which it ranked among the top five species on the basis of frequency of occurrence. It ranked number one in London, Paris, Sydney and Johannesburg. London plane tree's popularity as an urban tree is not limited to a single biome. The six cities in which it is among the top five species occur in the broadleaf evergreen forest, deciduous forest, and grassland biomes. Norway maple (*Acer platanoides*) ranked in the top five species in six cities as well: New York, Paris, Seattle, Moscow, Stockholm, and Vancouver. It is the species with the highest frequency of occurrence in Seattle. Like the London plane tree, Norway maple is dominant in more than one biome. Canary Island date palm was also among the top ranked species in six cities These cities occurred in the Mediterranean woodland/scrub, grassland, and savanna/scrub biomes. It is frequently used in the medians of boulevards because of its stately appearance and capacity to grow in a limited space.

Four other species, common lime (*Tilia europea*), silver birch (*Betula pendula*), Carolina poplar (*Populus x canadensis*), and Royal palm (*Oreodoxa regia*) each occurred among the top five species in four cities. The common lime (*Tilia europea*) is the most frequent species in Moscow and Stockholm. It also ranked among the top five species in London and Paris. It is a stately tree that has long been used in the urban areas in Europe. In northern European cities silver birch (*Betula pendula*) is also common. It is the top ranked species in Murmansk and it also occurs among the top five species in Stockholm, Reykjavik, and Seattle. It is an attractive tree with a capacity to withstand very low winter temperatures. Carolina poplar (*Populus x canadensis*) was found among the top five species in four cities: Moscow, Athens, Cusco, and Quito. It is a hybrid between North American cottonwood *(Populus trichocarpa)* and Europe's black poplar *(Populus nigra)* that was initially bred in the eighteenth century. Although its primary use has been as a forest plantation tree for lumber and wood pulp, it has found favor as a street tree due to its rapid growth and tolerance of urban conditions. Most of the cultivars of the hybrid used in cities are male trees to avoid problems associated with the cottony material surrounding the seed. Royal palm *(Oreodoxa regia)* is a common species in cities in the tropical forest, broadleaf evergreen forest, and desert biomes. It was among the top five species in Rio de Janeiro and Singapore, as well as in Sydney and Cairo. The presence of royal palm as a top ranked species in these cities is evidence of its popularity and its tolerance of the environmental constraints in three different biomes.

Street and Boulevard Trees

Streets and boulevards are urban environments where trees create shade, provide wind protection, generally improve conditions for human health and safety, and enhance aesthetic quality. Allan Jacobs has stated that trees are not required for "great streets", but most "great streets" are enhanced by the presence of trees (Jacobs 1993). Trees are often selected for planting along streets and boulevards on the basis of their form, their tolerance of restricted growing space, their vigor under the constraints of more intensively developed portions of urban areas, and their minimum maintenance requirements. Trees suitable for planting in parks and other urban environments may not be suitable for use as street and boulevard trees. As a result the list of the most frequently encountered urban trees, based on frequency of occurrence, is not the same as the list of most frequently encountered street trees. Only 75 of the 115 tree species listed in the table showing the five most common tree species in each city studied and the frequencies of their occurrence (Table 5.4) occurred among the top five street and boulevard trees in these same cities (Table 5.7). The top five street and boulevard tree species with the highest frequencies of occurrence includes 31 species that did not rank among the top five species in all environments. For example, in Los Angeles the top five species, based on their frequencies of occurrence in all environments, are Mexican fan palm (*Washingtonia robusta)*, Bailey acacia *(acacia bailyana)*, southern magnolia (*Magnolia grandiflora)*, Indian laurel fig *Ficus retusa)*, and Canary Island date palm (*Phoenix canariensis)*. Only three of these, Mexican fan palm, Canary Island date palm, and Indian laurel fig were ranked among the top five street and boulevard trees on the basis of frequency of occurrence.

A more detailed analysis of the top five street and boulevard trees encountered in the cities studied in each biome reveals some interesting patterns. In a given biome

Table 5.7 Top five street and boulevard trees on the basis of frequency of occurrence

Biome	City	Top five street and boulevard species		
		Common name	Scientific name	Freq.
Tropical forest	Bangkok	Angsana	*Pterocarpus indicus*	32
		Trumpet Tree	*Tabebuia rosea*	33
		Golden Shower	*Cassia fistula*	17
		Honduras Mahogany	*Swietenia macrophylla*	17
		Bullet Wood	*Minusops elengi*	17
	Rio de Janeiro	Amendoeira	*Lagerstroemia speciosa*	38
		Royal Palm	*Oreodoxa regia*	25
		Tipa Blanca	*Tipuana tipu*	16
		Morton Bay Fig	*Ficus macrophylla*	10
		Coconut Palm	*Cocos nucifera*	9
	Singapore	Angsana	*Pterocarpus indicus*	31
		Yellow Flame	*Peltophorum pterocarpum*	25
		Rubber Tree	*Ficus elastica*	13
		Tamarind	*Tamarindus indica*	12
		Royal Palm	*Oreodoxa regia*	12
Broadleaf evergreen forest	Hong Kong	Indian Laural Fig	*Ficus retusa*	25
		Candlenut Tree	*Aleurites moluccana*	20
		Silk Cotton Tree	*Bombax ceiba*	15
		Camphor Tree	*Cinnamomum camphora*	10
		Chinese Banyan	*Ficus benjamina*	10
	Sydney	London Plane Tree	*Platanus* x *hispanica*	21
		Brown Gum	*Eucalyptus robusta*	21
		Lemon Bottlebrush	*Callistemon lanceolatus*	14
		Rigid leaf Tea	*Melaleuca styphelioides*	14
		Flame Tree	*Sterculia acerifolia*	10
	Tokyo	Ginkgo	*Ginkgo biloba*	33
		Zelkova	*Zelkova serrata*	33
		Camphor Tree	*Cinnamomum camphora*	22
		Mochi Tree	*Ilex integra*	22
		Black Locust	*Robinia pseudoacacia*	11

(continued)

Table 5.7 (continued)

Biome	City	Top five street and boulevard species		
		Common name	Scientific name	Freq.
Deciduous forest	London	London Plane	*Platanus* x *hispanica*	67
		Common Lime	*Tilia europea*	20
		Rowan	*Sorbus aucuparia*	13
		English Elm	*Ulmus campestris*	7
		Field maple	*Acer campestre*	7
	New York	Honey Locust	*Gleditsis triacanthos*	46
		Norway Maple	*Acer platanoides*	31
		London Plane Tree	*Platanus* x *hispanica*	31
		Basswood	*Tilia americana*	23
		Northern Red Oak	*Quercus rubra*	23
	Paris	London Plane Tree	*Platanus* x *hispanica*	67
		Common Ash	*Franinus excelsior*	33
		Black Locust	*Robinia pseudoacacia*	20
		Horse Chestnut	*Aesculus carnea*	20
		Common Lime	*Tilia europea*	10
Coniferous forest	Jacksonville	Live Oak	*Quercus palustris*	70
		Laurel Oak	*Quercus laurifolia*	50
		Cabbage Palmetto	*Sabal palmetto*	30
		Crape-Myrtle	*Lagerstroemia indica*	20
		Bradford Pear	*Pyrus calleryana*	20
	Seattle	Basswood	*Tilia americana*	29
		American Elm	*Ulmus americana*	21
		Western Red Cedar	*Thuja plicata*	14
		Norway Maple	*Acer platanoides*	14
		Sugar Maple	*Acer saccharun*	14
	Vancouver	Japanese Flowering Cherry	*Prunus serrulata*	22
		Norway Maple	*Acer platanoides*	17
		London Plane Tree	*Platanus* x *hispanica*	11
		American elm	*Ulmus americana*	11
		Sugar maple	*Acer saccharum*	11

(continued)

Table 5.7 (continued)

Biome	City	Top five street and boulevard species Common name	Scientific name	Freq.
Mixed deciduous-coniferous forest	Beijing	Japanese Pagoda Tree	*Sophora japonica*	56
		Chinese White Poplar	*Populus tomentosa*	44
		Chinese Juniper	*Juniperus chinensis*	25
		Black Locust	*Robinia pseudoacacia*	19
		Chinese Red Pine	*Pinus tabulaeformis*	12
	Moscow	Common Lime	*Tilia europea*	83
		Norway Maple	*Acer platanoides*	43
		Silver Birch	*Betula pendula*	33
		Rowan	*Sorbus aucuparia*	25
		Carolina Poplar	*Populus x canadensis*	25
		Common Lime	*Tilia europea*	50
		Horse Chestnut	*Aesculus carnea*	20
	Stockholm	Common Lime	*Tilia europea*	50
		Horse Chestnut	*Aesculus carnea*	20
		Norway Maple	*Acer platanoides*	10
		Rowan	*Sorbus aucuparia*	10
		Dutch Elm	*Ulmus x hollandica*	10
Mediterranean woodland/scrub	Athens	White Mulberry	*Morus alba*	45
		Carolina Poplar	*Populus canadensis*	27
		Jacaranda	*Jacaranda mimosaefolia*	27
		Victorian-Box	*Pittosporum undulatum*	18
		Canary Island Date Palm	*Phoenix canariensis*	18
	Los Angeles	Indian Laural Fig	*Ficus retusa*	28
		Canary Island Date Palm	*Phoenix canariensis*	19
		Mexican Fan Palm	*Washingtonia robusta*	19
		Jacaranda	*Jacaranda mimosaefolia*	19
		Chinese Elm	*Ulmus parvifolia*	17
	Tel Aviv	Indian Laural Fig	*Ficus retusa*	28
		Jacaranda	*Jacaranda mimosifolia*	10
		Crape Myrtle	*Lagerstroemia indica*	5
		Mexican Fan Palm	*Washingtonia robusta*	3
		Canary Isalnd Date Palm	*Phoenix canariensis*	3

(continued)

Table 5.7 (continued)

Biome	City	Top five street and boulevard species		
		Common name	Scientific name	Freq.
Savanna/scrub	Brasilia	Ibira Pita	*Peltophorum dubium*	41
		Ceibo	*Erythrina crista-galli*	33
		Mahogany	*Swietenia macrophylla*	33
		Trumpet Tree	*Tabebuia chrysantha*	29
		White Orchid Tree	*Bauhinia forficata*	25
	Lagos	Umbrella tree	*Schefflera actinophylla*	42
		Royal Palm	*Oreodoxa regia*	25
		Australian beefwood	*Casuarina cunninghamiana*	21
		Canary Isalnd Date Palm	*Phoenix canariensis*	17
		Mast Tree	*Polyalthia longigolia* var. *pendula*	15
	New Delhi	Angsana	*Pterocarpus indicus*	27
		Pipal	*Ficus religiosa*	27
		Asoka	*Saraca indica*	18
		Babul	*Acacia nilotica*	18
		Neem	*Azadirachta indica*	9
Grassland	Buenos Aires	Green Ash	*Fraxinus pennsylvanica*	47
		London Plane Tree	*Platanus* x *hispanica*	40
		Paraiso	*Melia azedarack*	27
		Jacaranda	*Jacaranda mimosifolia*	27
		Black Locust	*Robinia pseudoacacia*	7
	Johannesburg	London Plane Tree	*Platanus* x *hispanica*	24
		Jacaranda	*Jacaranda mimosifolia*	24
		English Oak	*Quercus robur*	17
		Chinese Elm	*Ulmus parvifolia*	17
		Trident Maple	*Acer buergerianum*	10
	Tehran	Oriental Plane Tree	*Platanus orientalis*	54
		Eldarica Pine	*Pinus eldarica*	27
		Arizona Cypress	*Cupressus arizonica*	18
		Smooth Elm	*Ulmus carpinifolia*	9
		Persian Ash	*Fraxinus oxycarpa*	9

(continued)

Table 5.7 (continued)

Biome	City	Top five street and boulevard species		
		Common name	Scientific name	Freq.
Desert	Cairo	Indian Laurel Fig	*Ficus retusa*	73
		Nile Acacia	*Acacia nilotica*	60
		Palmyra Palm	*Borassus flabellifer*	17
		Royal Palm	*Oreodoxa regia*	17
		Rubber Tree	*Ficus elastica*	13
	Dubai	Canary Island Date Palm	*Phoenix canariensis*	67
		Banyan	*Ficus benghalensis*	17
		Royal Poinciana	*Delonix regia*	17
		Silk Tree	*Albizzia julibrissin*	17
		Neem	*Azadirachta indica*	17
	Phoenix	Mexican Fan Palm	*Washingtonia robusta*	45
		Palo Verde	*Cercidium torreyanum*	45
		Silver-dollar gum	*Eucalyptus polyanthemos*	27
		Canary Island Date Palm	*Phoenix canariensis*	27
		Arizona Ash	*Fraxinus velutina*	18
Tundra	Murmansk	Rowan	*Sorbus aucuparia*	100
		Downy Birch	*Betula pubescens*	60
		European Aspen	*Populus tremula*	20
		Siberian Mountain Ash	*Sorbus gorodkowii*	20
		Siberian Larch	*Larix sibirica*	20
	Nuuk	(no street trees)		
	Reykjavik	Rowan	*Sorbus aucuparia*	67
		Downy Birch	*Betula pubescens*	50
		Black Cottonwood	*Populus trichocarpa*	33
		European Larch	*Larix decidua*	17
		Gray Alder	*Alnus incana*	17

(continued)

Table 5.7 (continued)

		Top five street and boulevard species		
Biome	City	Common name	Scientific name	Freq.
Highland	Cusco	Monterey pine	*Pinus radiata*	43
		Carolina poplar	*Populus* x *canadensis*	28
		Chinese toon	*Cedrela sinensis*	28
		Silver wattle	*Acacia dealbata*	14
		Australian beefwood	*Casuarina cunninghamiana*	14
	Lhasa	Chinese White Poplar	*Populus tomentosa*	70
		Chinese Red Pine	*Pinus tabulaeformis*	30
		Chinese Juniper	*Juniperus chinensis*	20
		Weeping Willow	*Salix babylonica*	10
		Siberian Elm	*Ulmus pumila*	10
	Quito	Blackwood acacia	*Acacia melanoxylon*	29
		Carolina poplar	*Populus* x *canadensis*	29
		Black Elderberry	*Sambucus nigra*	24
		Dealbata acacia	*Acacia dealbata*	14
		White Poplar	*Populus alba*	14

one may observe some of the same species among the top five street and boulevard trees in each city studied in that biome. For example, the London plane tree was among the five dominant trees in each of the three cities in the deciduous forest biome (Table 5.8). Common lime and honey locust were also found in two of the three cities in that biome. The common use of these three species in the deciduous forest biome may be related to the geographical proximity of London and Paris and to the historical development of urban forests in Europe and North America. In contrast, no species occurring among the top five street and boulevard trees in each of the three cities in the savanna/scrub biome was common to all three cites (Brasilia, Lagos, New Delhi). The lack of common use of a species in these three cities may be due to their geographic separation, each occurring on different continents and to cultural differences among the three regions in which the cities occur.

Common lime, Norway maple, and rowan were among the top five species of street and boulevard trees in two cities (Moscow, Stockholm) in the coniferous forest biome, but not Jacksonville. The warmer climate of Jacksonville, as compared to Seattle and Vancouver, is in large part responsible for the absence of the three species common to Seattle and Vancouver.

In Jacksonville three native species [live oak (*Quercus palustris*), laurel oak (*Quercus laurifolia*), and cabbage palmetto (*Sabal palmetto*) and two exotic species (crape-myrtle (*Lagerstroemia indica*), Bradford pear (*Pyrus calleryana*)] are commonly used for street and boulevard planting. The three native species utilized in

Table 5.8 Comparison of the use of street and boulevard trees in cities in each biome

Biome/cities	Species in common			
Tropical forest	Angsana		Royal Palm	
	Pterocarpus indicus		*Oreodoxa regia*	
Bangkok	•			
Rio de Janeiro			•	
Singapore	•		•	
Broadleaf evergreen forest	Camphor Tree			
	Cinnamomum camphora			
Hong Kong	•			
Sydney	No species common with Hong Kong and Tokyo			
Toyko	•			
Deciduous forest	London Plane		Common Lime	
	Platanus x *hispanica*		*Tilia europea*	
London	•		•	
New York	•		•	
Paris	•			
Coniferous forest	American Elm	Norway Maple	Sugar Maple	
	Ulmus americana	*Acer platanoides*	*Acer saccharum*	
Jacksonville	No species common with Seattle and Vancouver			
Seattle	•	•	•	
Vancouver	•	•	•	
Mixed deciduous-coniferous forest	Norway Maple	Common Lime	Rowan	
	Acer platanoides	*Tilia europea*	*Sorbus aucuparia*	
Moscow	•	•	•	
Stockholm	•	•	•	
Beijing	No species common with Moscow and Stockholm			
Mediterranean woodland/scrub	Jacaranda	Indian Laurel Fig	Mexican Fan Palm	Canary Island
	Jacaranda mimosifolia	*Ficus retusa*	*Washingtonia robusta*	Date Palm
				Phoenix canariensis
Athens	•			•
Los Angeles	•	•	•	•
Tel Aviv	•	•	•	•
Savanna/scrub	No common street or boulevard species			
Grassland	London Plane		Jacaranda	
	Platanus x *hispanica*		*Jacaranda mimosifolia*	
Buenos Aires	•		•	
Johannesburg	•		•	

(continued)

Table 5.8 (continued)

Biome/cities	Species in common	
Tehran	No species common with Buenos Aires and Johannesburg	
Desert	Canary Island Date Palm	
	Phoenix canariensis	
Cairo	•	
Dubai	•	
Phoenix	•	
Tundra	Rowan	Downy Birch
	Sorbus aucuparia	*Betula pubescens*
Murmansk	•	•
Reykjavik	•	•
Nuuk	No street or boulevard trees	
Highlands	Carolina Poplar	
	Populus x *canadensis*	
Cusco	•	
Lhasa	No species common with Cusco and Quito	
Quito	•	

• means the species is present

Jacksonville are v well adapted to the colder climate of Seattle and Vancouver where more northern species [basswood (*Tilia americana)*, American elm (*Ulmus americana)*, western red cedar (*Thuja plicata)*, sugar maple (*Acer saccharum)*, Norway maple (*Acer platanoides*)] are employed. With the exception of western red cedar, the more commonly observed street and boulevard trees used in Seattle and Vancouver are not native to the region or to the coniferous forest biome. The general absence of conifer species as street and boulevard trees in the cities studied in the coniferous biome may be due to the large size at maturity of the local conifers species [Jacksonville: longleaf pine (*Pinus palustris*), bald cypress *(Taxodium distichum)*, loblolly pine (*Pinus taeda*); Seattle and Vancouver: Douglas-fir *(Pseudotsuga menziesii*), western hemlock *(Tsuga heterophylla),* Sitka spruce *(Picea sitchensis*)] or to the desire for visual alternatives to the uniformity of the coniferous forest. Street and boulevard environments often do not have the space necessary for such large trees. The widespread use of hardwood species for street and boulevard trees in the coniferous forest biome adds more color to the urban forest by introducing flowering trees (e.g., crape-myrtle and Bradford pear in Jacksonville; Japanese flowering cherry in Vancouver) or trees noted for their lighter green summer foliage and their fall colors (sugar maple, Norway maple, American elm, and basswood in Seattle and Vancouver).

The use of different tree species for street vs. boulevard planting was not often significant in the cities studied. Median strips in boulevards often provide more growing space than is provided along the edges of boulevards or along city streets. In some cities (e.g., Dubai, Cairo, Athens) the medians were planted with palms, while other species tended to be used along streets. However, many cities employed the same tree species in the planting of both streets and boulevards. Some cities had few if any boulevards divided with median strips (e.g., London, Jacksonville, Reykjavik). The number of boulevards examined was usually too small to allow for any definitive comparison to streets in most of the cities studied. Therefore, the data on tree species observed along streets and boulevards were combined to give a larger data set upon which to base conclusions.

Traffic Barriers

Trees are used in some cities as traffic barriers to slow or prohibit vehicular traffic. A commonly used structure is the traffic circle. The raised central portion of a traffic circle requires vehicles entering an intersection to slow down and proceed around a portion of the circle before continuing along the initial direction of travel or turning into the intersecting street. Traffic circles reduce the speed of vehicle, and also reduce the number of collisions at intersections. Trees are planted in traffic circles to increase awareness of a major traffic intersection and to improve the visual quality of these intersections.

Traffic diversion is achieved in some cities by closing off a section of a street and planting trees in the closed-off section. This prevents traffic from entering the street and prevents the street from being used as a thoroughfare. Such partial street closures have successfully quieted traffic in residential neighborhoods (Appleyard et al. 1981). Trees that are incorporated into these closed off sections of streets serve to indicate to the driver that passage along the street is blocked. They also provide a park-like setting for the neighborhood where a street has been blocked.

Tree species richness in the large traffic circles observed in Dubai, New Delhi, and Quito varied from one to three species (Table 5.9). Eighty percent of the traffic circles in Dubai used only one species, in Quito 50%, and in New Delhi 37.5%. Fifty percent of the traffic circles in New Delhi supported two tree species, while in Dubai and Quito the percentages were 20 and 25 respectively. Three species were used in 25% of the traffic circles in Dubai and in 12.5% in New Delhi. No traffic circles in Quito used three species. None of the traffic circles in the three cities supported more than three species. There was considerable variation in the tree species used in traffic circles in each of the cities. No species was common neither to all three cities nor to two of the three cities. Palms were, however, frequently employed in both Dubai (Mexican fan palm *Washington robusta*) and New Delhi (Royal palm *Roystonea regia*).

The small traffic circles observed in Seattle supported an average of 1.4 trees with a range of one to four. Oblong traffic barrier in Seattle supported an average of 1.6 trees, with a modal number of one. Tree canopy cover averaged only 15%. A total of 26 tree species were observed on the traffic circles and oblong barriers in Seattle. However, species richness ranged from one to four species for both the traffic circles and the oblong traffic barriers. The modal number of species used was only one, 88.6% of the traffic circles and oblong barriers supported only one tree species.

Table 5.9 Characteristics of traffic circles

	City					
Characteristic	Dubai	New Delhi	Quito	Tel Aviv	Seattle	Vancouver
Modal diameter (ft.)	200	200	200	100	15	16
Range in Diameter (ft.)	100 to 200	100 to 200	140 to 200	90 to 100	10 to 24	14 to 20
Modal tree number	8	10	8	1	1	1
Range in tree numbers	1 to 18	3 to 34	7 to 8	1 to 5	1 to 4	1 to 2
Number of tree species	5	12	6	3	14	6
Most common species	Mexican Fan Palm	Ashoka	Dragon Tree	Canary Island Date Palm	Japanese Flowering Cherry	Japanese Flowering Cherry
	Washingtonia robusta	*Saraca indica*	*Dracaena draco*	*Phoenix canariensis*	*Prunus serrulata*	*Prunus serrulata*
Ave. canopy cover (%)	11	21	10	10	20	20

The most commonly encounter species were flowering cherry (*Prunus serrulata*) (occurring on 27.5% of the sites), Norway maple (*Acer platanoides*) (24.1%), silver birch (*Betula pendula*) (13.8%), and crabapple (*Malus sp.*) (13.8%). Three of these species (flowering cherry, silver birch, and crabapple) are small stature trees at maturity and were selected due to the limited growing space.

Traffic diverters were created in Sydney by closing off a street at an intersection thus preventing access to the closed off street and the use of that street by through traffic. Seven tree species were recorded on traffic diverters in Sydney. These were London plane (*Platanus x acerifolia*), jacaranda (*Jacaranda mimosifolia*), zelkova (*Zelkova serrata*), catalpa (*Catalpa speciosa*), green ash (*Fraxinus pennsylvanica*), royal palm (*Roystonea regia*), and brown gum (*Eucalyptus robusta*). With the exception of London plane, these species had a low frequency of occurrence. The frequency of occurrence of London plane was 60%, jacaranda 40%, and the remaining species only 20% each. On average only two tree species were planted in each of the traffic diverters surveyed in Sydney.

Parks

Urban parks are the places where most urban dwellers encounter trees. Although large parks usually support a large variety of tree species, it is the small urban park where the average person experiences trees on a regular basis. Here people often learn the few trees aside from street trees that they associate with the neighborhood. Small parks were the focus of this study in part because they were expected to better represent an expression of the local use of trees in park in each of the biomes. It was also thought that they might better represent local cultural and historical variations since small parks were less likely to be the products of national or international design competitions. Trees in small parks could also be inventoried relatively quickly in comparison to the time required for large urban parks. The parks that are reported here were from one to two acres in size, often occupying a city block. Some were a bit smaller, as little as 100 feet by 100 feet, but still large enough to accommodate some trees.

A total of 221 tree species were encountered in the 157 small parks sampled. The number of species encountered in small parks ranged from one to eighteen, with an average number of 6 tree species (Table 5.10). The average number of species used in small parks in each of the biomes ranged from 3.9 to 7.1 (Table 5.11). The highest average number of species in small parks occurred in cities in the savanna/ scrub and the lowest was in the coniferous forest biome. The variation in the average number of species did not follow any particular pattern among the biomes. The average for the forested biomes was 5.2 while the average for the non-forested biomes was 6.2.

Seventeen tree species occurred in small parks in five or more cities (Table 5.12). The most common of these were black locust (11 cities), London plane tree (8 parks), Canary Island date palm (8 cities), and southern magnolia (7 cities). The first four of these are among the 10 most frequently encountered street and boulevard trees. The common use of these species is due to their adaptation to urban environments across a number of biomes. The common occurrence of black locust and southern magnolia in small parks may be also due to their attractive flowers and foliage. They were used as specimen trees in many of the parks. Canary Island date palm was used as a specimen tree and also used to line walks through parks in warmer non-forested biomes. The London plane tree was commonly observed as a tree lining walks or along the perimeters of small parks.

No attempts were made to record the large number of tree species occurring in the large parks in the cities studied. Many of these parks were reported to have over 150 species (Fabos et al. 1968; Montero 2001). Large parks were visited in each of the cities studied to obtain information on their origin, design, and general tree cover. During these large park surveys general notes were made on the species composition of each park. It was evident that large urban parks in all biomes rely on native tree species to a large extent. This may reflect the availability of planting stock at the time many of these parks were established. Although a dependence on trees native to the biome and geographical area in which the city was located was

Table 5.10 Species present in small parks

Biome	TF			BEF			DF			CF			MDCF			MWS			SS			G			D			T			H			SUM
City	1	2	3	4	5	6	7	8	9	10	11	12	13	14	15	16	17	18	19	20	21	22	23	24	25	26	27	28	29	30	31	32	33	
Species																																		
Acacia delbata																									•							•		2
Acacia farnesiana																																•		1
Acacia melanoxylon																	•															•		2
Acacia pycnantha																											•							1
Acacia saligna																											•							1
Acacia xarthophloca																			•															1
Acer buergerianum							•							•									•											3
Acer campestre																													•					1
Acer macrophyllum												•		•																				2
Acer negundo														•		•							•											3
Acer palmatum					•		•					•																						3
Acer platanoides								•				•			•								•											4
Acer pseudoplatanus							•				•																							2
Acer pycnanthum					•																													1
Acer rubra												•																						1
Acer saccharinum							•					•																						2
Acer saccharum								•			•	•																						3
Acer saccharum var. florida										•																								1
Acer tataricum													•																					1
Aesculus hippocastanum							•	•	•			•		•	•																			6
Ailanthus altissima			•				•																										•	3

(continued)

Table 5.10 (continued)

Biome	TF			BEF			DF			CF			MDCF			MWS			SS			G			D			T			H			SUM
Albizia julibrissima	•					•											•						•	•								•		6
Albizia lebbeck																										•								1
Alnus acuminata																																•		1
Alnus incana																													•	•				2
Alnus rubra												•																						1
Alstonia religiosa																				•														1
Andira inermis			•																															1
Araucaria araucana																																	•	1
Araucaria angustifolia																						•												1
Araucaria heterophylla																		•														•		2
Areca catechu						•																												1
Azadirachta indica																				•			•		•	•								4
Balsamodendrum roxburghii																				•														1
Baubuia forficata																			•															1
Bauhinia purpurea						•																												1
Betula papyrifera												•																						1
Betula pendula								•				•		•															•					4
Betula pubescens																												•	•					2
Betula subarctic																												•						1
Bombax ceiba						•														•														2
Borassus flabellifer																				•														1
Brugmansia arborea																																	•	1
Buddleia purpurea																							•											1
Callistemon citrinus																	•	•														•	•	4

Callistemon salignus																															•		1
Cassia fistula						•											•			•													3
Cassia siamea																																•	1
Cassia surattensis						•																											1
Cassia tomentosa																															•		1
Casuarina equisetifolia			•	•		•											•		•														5
Casuarina stricta																															•		1
Cecropia insignis		•																															1
Cedrela fissilis																																•	1
Cedrela sinensis																																•	1
Cedrus atlantica										•	•														•								3
Cedrus deodara					•						•					•					•	•	•							•			7
Cedrus libani																	•																1
Ceiba speciosa																		•			•												2
Ceiba pentandra																	•																1
Celtis australis					•										•			•				•											4
Celtis occidentalis								•			•																						2
Cericidium floridum																										•							1
Cercis canadensis								•																									1
Cercis siliquastrum							•																•										2
Chamaecyparis lawsoniana											•																						1
Chamaecyparis nootkatensis											•																						1

(continued)

Table 5.10 (continued)

Biome	TF	BEF	DF	CF	MDCF	MWS	SS	G	D	T	H	SUM
Chamaecyparis obtusa				•								1
Chilopsis linearis									•			1
Cinnamomum camphora						•						1
Citrus aurantifolia						• •	•					3
Citrus sinensis						•			•			2
Cocos nucifera	• •								•		•	4
Commiphora neglecta						•						1
Cornus alternifolia				•								1
Cornus kousa		•										1
Cornus nuttallii				•								1
Cotoneaster frigidus				•								1
Crataegus monogyna			•		•			•				3
Crataegus oxyacantha					• •			•		•		4
Crataeva religiosa	•											1
Crataeva unilocularis		•										1
Cryptomeria japonica		•		•			•					3
Cupressus arizonica						•		•				2
Cupressus macrocarpa								•			•	2
Cupressus sempervirens						•			•			1
Cupressus semp. 'fascata'								•				1
Delonix regia	• •	•				•	•	•	• • •			9
Dictyosperma album		•										1
Dracaena draco		•										1
Elaeagnus angustifolia											•	1

Enterolobium contortisiliquum																			•															1
Eriobotrya japonica																		•														•		2
Erythrina cristagalli			•														•		•														•	4
Eucalyptus camaldulensis							•																		•		•							3
Eucalyptus cinerea																											•							1
Eucalyptus citridora						•												•																2
Eucalyptus globulus																																•		1
Eucalyptus robusta				•																														1
Eucalyptus sideroxylon																																•		1
Eugenia jumblana																										•								1
Fagracea crenulata			•															•																2
Fagus americana								•																										1
Fagus sylvatica							•					•																						2
Ficus benghalensis		•																			•					•								3
Ficus benjamina						•																												1
Ficus carica																	•	•					•											3
Ficus elastica		•	•															•				•			•									5
Ficus lutea																			•															1
Ficus macrocarpa		•																•							•	•								4
Ficus macrophylla		•		•													•																	3
Ficus religiosa		•																•		•														3
Ficus sycomorus																		•							•									2

(continued)

Table 5.10 (continued)

Biome	TF	BEF	DF	CF	MDCF	MWS	SS	G	D	T	H	SUM
Ficus vilifolia	•											1
Ficus virens		•										1
Fraxinus excelsior					•			•			•	3
Fraxinus oxycarpa var. oxycarpa								•				1
Fraxinus pennsylvanica			• •	• •				•				5
Fraxinus uhdei						•						1
Fraxinus velutina									•			1
Gaptodea camped	•											1
Geijera parviflora								•				1
Gingko biloba		•										1
Gleditsia triacanthos		•	•								•	3
Grevilla robusta		•				•		•				3
Gymnocladus dioica					•							1
Ilex aquifolium			•	•								2
Jacaranda mimosifolia						•		• •				3
Juglans californica									•			1
Juglans neotropica											•	1
Juglans regia			•	•								2
Juniperus chinensis		•		•			•				•	4
Juniperus virginiana				•								1
Kigelia pinnata	•											1
Lagerstroemia indica	•	•		•			•					4
Lagerstroemia speciosa	•						•					2

																																		I
Larix decidua														•																				1
Larix occidentalis												•																						1
Leptospermum laevigatum																																	•	1
Liquidambar formosana												•																						1
Liquidambar styraciflua										•	•	•				•	•																	5
Liriodendron tulipifera							•					•											•											3
Livistona chinensis						•																												1
Maclura pomifera								•																										1
Magnoila campbellii												•																						1
Magnolia grandiflora										•	•					•	•	•				•		•								•		7
Magnolia kobus					•																													1
Magnolia soulangeana							•					•																						2
Malus domestica														•									•											2
Malus sylvestris							•	•																										2
Michella champaca																				•														1
Melaleuca leucadendron																			•															1
Melia agedarach																			•			•												2
Metasequoia glyptostroboides					•																													1
Morus alba																•		•																2
Neodypsis decary						•																												1
Nyssa aquatica										•																								1
Olea europea																		•					•											2
Oreodoxa regia	•	•																	•	•			•		•									6

(continued)

Table 5.10 (continued)

Biome	TF			BEF			DF			CF			MDCF			MWS			SS			G			D			T			H			SUM
Parkinsonia aculeata																				•													•	2
Paulonia tomentosa							•																											1
Peltophorum pterocarpum	•		•																•															3
Pheonix canariensis										•						•	•	•	•			•	•			•								8
Phoenix dactylifera																		•																1
Picea abies														•																				1
Picea aurantiaca																															•			1
Picea pungens																								•										1
Picea sitchensis												•																						1
Pinus canariensis																	•																	1
Pinus contorta												•																						1
Pinus densiflora												•																						1
Pinus halepensis																											•							1
Pinus nigra												•																						1
Pinus palustris										•																								1
Pinus pinea																	•																	1
Pinus ponderosa												•																						1
Pinus sylvestris											•																							1
Pinus radiata																							•									•	•	3
Pinus eldarica var. eldarica																								•										1
Pinus taeda										•																								1
Pinus tabuliformis																															•			1
Pinus wallichiana																															•			1
Pithecallobium duice																				•														1

Pittosporum undulatum				•												•	•																	3
Platanus occidentalis										•																								1
Platanus orientalis					•																			•	•									3
Platanus x *hispanica*				•			•	•	•			•				•	•					•	•											9
Platycaldus oreintalis																															•			1
Plumeria acuminata	•																																	1
Plumeria rubra						•																												1
Podocarpus macrophyllus						•																												1
Polyalthia longifolia																					•													1
Populus alba																						•		•								•		3
Populus nigra																								•										1
Populus nigra 'Italica'															•	•			•			•		•										5
Populus tomentosa													•																		•			2
Populus tremula														•														•						2
Populus tristis																												•						1
Populus trichocarpa											•																							1
Populus x *canadensis*																•						•										•		3
Prosopis alba																			•															1
Prosopis pubescens																											•							1
Prosopis velutina																											•							1
Prunus avium							•																											1
Prunus cerasifera							•					•											•								•			4
Prunus fastigate														•																				1

(continued)

Table 5.10 (continued)

Biome	TF			BEF			DF			CF			MDCF			MWS			SS			G			D			T			H			SUM
Prunus paduas															•													•						2
Prunus serrulata									•			•	•		•								•								•			6
Prunus x *yedoensis*					•																													1
Pseudotsuga menziesii												•																						1
Pterocarpus indicus	•		•																•															3
Punica granatum																								•										1
Putranjiva roxburghii																				•														1
Pyrus calleryana										•																								1
Quercus kunugi					•																													1
Quercus cerris										•																								1
Quercus ilex							•									•							•											3
Quercus laevis												•																						1
Quercus laurifolia										•																								1
Quercus palustris								•				•										•												3
Quercus phellos																	•	•																2
Quercus prinus																							•											1
Quercus robur							•							•								•	•											4
Quercus rubra																							•											1
Quercus velutina								•		•		•																						3
Quercus virginiana										•																								1
Ravenala madagascariensis						•																												1
Rhododendron arboreum												•																						1
Robinia pseudoacacia					•		•		•			•				•	•					•	•	•	•							•		11
Roystonea regia						•																												1

Sabal palmetto									•																									1
Salix alba													•										•											2
Salix babylonica	•					•					•										•		•							•				6
Salix caprea							•							•																				2
Salix daphnoides																														•				1
Salix fragilis									•																									1
Salix glauca																													•					1
Salix shidare					•																													1
Salix tetrasperma																			•											•				2
Salix vuninalis																														•				1
Sambucus nigra														•																	•			2
Sapium sebiferum																	•																	1
Saraca asoca																			•															1
Scheffera purkleri		•																																1
Schinus molle																										•						•		2
Senna splendida																		•																1
Sequoiadendron giganteum											•																							1
Sophora japonica												•																						1
Sorbus acuparius													•														•	•						3
Sorbus intermedia							•																											1
Sorbus torminalis																												•						1
Spathodea campanulata																		•																1
Sterculia foetida		•	•																															2

(continued)

Table 5.10 (continued)

Biome	TF	BEF	DF	CF	MDCF	MWS	SS	G	D	T	H	SUM
Swietenia macrophylla	•						•					2
Swietenia mahagoni		•										1
Syzygium panicultatum			•									1
Tabebuia chrysantha	•											1
Tabebuia rosea								•				1
Tamarindus indica	•											1
Tamarix aphylla						•						1
Taxodium distichum		•		•								2
Taxus baccata			•									1
Taxus brevifolia				•								1
Tectona grandis	•											1
Terminalia arjuna							•					1
Thuja plicata				•								1
Tilia americana			•									1
Tilia cordata			•	•	• •							4
Tilia x europaea			• •									2
Tipuana tipu	•						•	•				3
Toona sinensis			•									1
Trachicarpus fortunei											• •	2
Turraea villosa							•					1
Ulmus americana			•	•								2
Ulmus glabra var. *camperdowny*								•				1
Ulmus procera				•	•							2
Ulmus x hollandrica					•							1

Ulmus parvifolia																							•											1
Ulmus pumila																															•			1
Washingtonia fiilfera																	•					•	•		•	•	•							6
Washngtonia robusta												•																						1
Zelkovia serrata					•																													1
Ziziphus mauritiana																				•														1
Ziziphus spina-christi																										•								1
Total species = 284																																		
Sum	15	12	12	5	7	26	28	18	5	22	8	53	5	19	11	12	20	30	20	20	4	20	32	19	12	11	15	7	6	2	15	24	14	
City	1	2	3	4	5	6	7	8	9	10	11	12	13	14	15	16	17	18	19	20	21	22	23	24	25	26	27	28	29	30	31	32	33	
Biome	TF			BLEF			DF			CF			MDCF			MWS			SS			G			D			T			H			

• means the species is present

1 = Bangkok; 2 = Rio de Janeiro; 3 = Singapore; 4 = Sydney; 5 = Tokyo; 6 = Hong Kong; 7 = London; 8 = New York; 9 = Paris; 10 = Jacksonville; 11 = Seattle; 12 = Vancouver; 13 = Beijing; 14 = Moscow; 15 = Stockholm; 16 = Athens; 17 = Los Angeles; 18 = Tel Aviv; 19 = Brasilia; 20 = New Delhi; 21 = Lagos; 22 = Buenos Aires; 23 = Johannesburg; 24 = Tehran; 25 = Cairo; 26 = Dubai; 27 = Phoenix; 28 = Murmansk; 29 = Reykjavik; 30 = Nuuk; 31 = Lhasa; 32 = Quito; 33 = Cusco

Table 5.11 Average number of species observed in small parks

Biome	No. of parks surveyed	Average no. species
Tropical forests	16	4.4
Broadleaf evergreen forest	15	6.5
Deciduous forest	16	6.2
Conifer forest	17	7.1
Mixed deciduous-coniferous forest	14	3.9
Mediterranean woodland/scrub	19	6.9
Savanna-thorn scrub	9	6.6
Grassland	16	7.1
Desert	16	5.0
Tundra	6	4.0
Highland	19	5.7

evident, many large parks have used large numbers of exotic tree species. Large parks in cities in the non-arboreal biomes were, of course, dependent on exotic species. Observations concerning the history and layout of the large parks surveyed are presented in the chapter on design.

Greenbelts

Green belts have been established at the margins of six of the cities included in this study. They serve a number of function including recreation, wildlife habitat, production of forest products, control of wind erosion, reduction of air pollution, and areas for the tertiary treatment of sewage water. The greenbelts surveyed in cities in forested biomes (Stockholm and Seattle) were dominated by tree species native to the local biome. These greenbelts consisted of areas of natural forest, although most of these were forests that had been previously harvested. Portions of the natural forest greenbelts are being used for commercial forest production and will continue to be harvested. In the greenbelts around Stockholm the commonly observed species include English oak (*Quercus robur*), Norway spruce (*Picea abies*), Norway maple (*Acer platanoides*), rowan (*Sorbus aucuparia*), Scots pine (*Pinus sylvestris*), and silver birch (*Betula pendula*). Greenbelts in Seattle were delineated in second growth forests dominated by Douglas-fir (*Pseudotsuga menziesii*), Western hemlock (*Tsuga heterophylla*), Western red cedar (*Thuja plicata*) Sitka spruce (*Picea sitchensis*), and Red alder (*Alnus rubra*) which flanked the city.

In non-forested biomes greenbelts are often planted. The selection of species often depends on the primary function of the greenbelt and the ability of species being considered to thrive in the local climate. Greenbelts have been established in the natural grasslands around Tehran by the planting of Afghan pine (*Pinus eldarica* var. *eldarica*) and Arizona cypress (*Cupressus arizonica*). These greenbelts produce pole sized material that are harvest for construction material and used Tehran. The

Table 5.12 Number of Species observed in small parks

Biome	City	Site #	No. species
Tropical forest	Bangkok	3	8
		8	1
		10	8
		17	2
	Number of Parks/Sum	(4)	19
	Ave		4.8
	Rio de Janeiro	7	4
		11A	6
		12A	1
		14A	2
		17	3
		18	1
		21	6
		22	6
	Number of parks/sum	(8)	29
	Ave		3.6
	Singapore	1	2
		5	5
		22	5
		30	7
	Number of Parks/Sum	(4)	19
	Ave		4.8
Broadleaf evergreen forest	Sydney	1	5
		2	5
	Number of Parks/Sum	(2)	10
	Ave		5
	Tokyo	9	4
		10	8
		11	8
		12	11
		28	2
	Number of Parks/Sum	(5)	33
	Ave		6.6
	Hong Kong	14	4
		34	21
		4	7
		23	8
		24	4
		26	9
		42	8
		44	3
	Number of Parks/Sum	(8)	64
	Ave		8

(continued)

Table 5.12 (continued)

Deciduous forest	London	2	12
		6	14
		7	5
		12	6
		17	7
		20	4
		22	10
		24	7
		26	4
		32	1
	Number of Parks/Sum	(10)	70
	Ave		7.0
	New York	9	2
		15	18
	Number of Parks/Sum	(2)	20
	Ave		10
	Paris	4	2
		6	2
		12	2
		25	3
	Number of Parks/Sum	(4)	9
	Ave		2.2
Coniferous forest	Jacksonville	4	4
		9	5
		15	8
		17	4
		24	8
		29	5
	Number of Parks/Sum	(6)	34
	Ave		5.7
	Seattle	23	6
		35	1
		36	1
		3	4
	Number of Parks/Sum	(4)	12
	Ave		3.0
	Vancouver	3	8
		4	9
		5	11
		6	12
		7	6
		9	11
		10	18
	Number of Parks/Sum	(7)	75
	Ave		10.7

(continued)

Table 5.12 (continued)

Mixed deciduous-coniferous forest	Beijing	A	5
		B	4
	Number of Parks/Sum	(2)	9
	Ave		4.5
	Moscow	1	9
		6	5
		21	2
		16	0
		17	2
		18	1
		D	5
		E	1
	Number of Parks/Sum	(8)	25
	Ave		3.6
	Stockholm	2	9
		11	3
		13	1
		17	1
	Number of Parks/Sum	(4)	14
	Ave		3.5
Mediterranean woodland/scrub	Athens	2	4
		17	9
		22	6
	Number of Parks/Sum	(3)	19
	Ave		6.3
	Los Angeles	4	8
		5	7
		22	3
		28	5
		30	10
	Number of Parks/Sum	(5)	33
	Ave		6.6
	Tel Aviv	7	8
		10	4
		11	5
		15	11
		15A	7
		20	3
		25	11
		39	4
		40	9
		33	13
		36	4
	Number of Parks/Sum	(11)	79
	Ave		7.1

(continued)

Table 5.12 (continued)

Savanna/scrub	Brasilia	4	4
		8	9
		10	8
	Number of Parks/Sum	(3)	21
	Ave		7.0
	Lagos	20	1
		31	3
		34	2
	Number of Parks/Sum	(3)	6
	Ave		2.0
	New Delhi	8	20
		18	12
		23	1
	Number of Parks/Sum	(3)	33
	Ave		11
Grassland	Buenos Aires	2	1
		4	5
		8	8
		22	5
		24	5
		27	6
	Number of Parks/Sum	(6)	30
	Ave		5.0
	Johannesburg	3	10
		13	5
		18	6
		21	4
		27	4
		33	9
		43	8
	Number of Parks/Sum	(7)	42
	Ave		6.0
	Tehran	1	15
		2	11
		3	15
	Number of Parks/Sum	(3)	41
	Ave		13.7

(continued)

Table 5.12 (continued)

Desert	Cairo	7	5
		12	5
		16	5
		33	13
	Number of Parks/Sum	(4)	28
	Ave		7.0
	Dubai	1	2
		2	7
		13	6
		17	3
		18	3
		25	2
		32	5
	Number of Parks/Sum	(7)	28
	Ave		4.0
	Phoenix	6	4
		11	3
		16	6
		33	8
		34	3
	Number of Parks/Sum	(5)	24
	Ave		4.8
Tundra	Murmansk	4	6
		5	4
		11	8
	Number of Parks/Sum	(3)	18
	Ave		6.0
	Nuuk	1	2
	Number of Parks/Sum	(1)	2
	Ave		2
	Reykjavik	7	1
	Number of Parks/Sum	(2)	4
	Ave		2.0

(continued)

Table 5.12 (continued)

Highland	Quito	6	9
		19	9
		23	2
		24	10
		28	6
		35	3
		38	3
		53	2
	Number of Parks/Sum	(8)	44
	Ave		5.5
	Lhasa	8	5
		9	7
		10	3
		13	4
		17	8
	Number of Parks/Sum	(5)	27
	Ave		5.4
	Cusco	2A	0
		4	3
		5	4
		10	2
		16C	5
		34	5
	Number of Parks/Sum	(6)	19
	Ave		3.2
Overall Average Number of Tree Species per Park			5.4

extensive greenbelts around the city of Beijing have been planted on agricultural land which thousands of years earlier supported mixed deciduous-conifer forests. In the Beijing greenbelts Chinese white poplar has been extensively planted because of its rapid growth rate, it adaptation to the climate of the Beijing area, its capacity to reduce wind velocity, and its ability to reduce the concentration of particulate matter in the air. Experimental greenbelts are being established near Cairo to serve as areas for tertiary treatment of sewage water. Water from partially treated sewage is used to irrigate species that are tolerant of the desert heat and the high mineral concentration of the water used for irrigation. These species included Arabian acacia (*Acacia farnesiana*), Arizona cypress (*Cupressus arizonica*), casuarina (*Casuarina cunninghamiana*), and red gum (*Eucalyptus camaldulensis*).

Waste Areas

Untended sites resulting from absentee ownership, demolition of buildings, urban clearance, or lack of maintenance occurs in many cities of the world. These sites are commonly called waste areas and they experience a natural establishment of vegetation, including trees. The vegetation of waste areas has been the focus of study in a number of European cities (Sukoop 2002). Much attention was directed by ecologists to the colonization of bombed sites in cities during World War II (Salisbury 1943; Lousley 1944; Scholz 1960). Invasion of bombed sites and sites resulting from the demolition of buildings, in which considerable building rubble is left on the site, is enhanced by wind dispersed of seeds and the ability of trees to survive under drier and warmer conditions. Wind dispersal often characterizes the early invaders of other urban waste areas, but variations in soil moisture can play an important role in the successful establishment of ruderal vegetation on these sites. Several exotic tree species introduced into urban forests and are known for their invasive behavior, are often found among the early colonizers of waste areas. The tree of heaven *(Ailanthus altissima)*, for example, is an aggressive invader of waste areas in many American cities (Plotkin 2000).

A total of 54 species were observed colonizing waste areas in the cities surveyed in this study (Table 5.13). The greatest number of tree species invading waste areas was recorded in the mixed deciduous-conifer forest biome, while the least numbers were recorded in the Mediterranean woodland scrub and grassland biomes. Stockholm exhibited the greatest number of species invading waste areas with a total of nine species, while no invasive tree species were observed on waste areas in Tehran or Cairo. Tree-dominated biomes had an average of four tree species recorded on waste areas, while an average of only two species were observed in non-tree, dominated biomes. This difference is due to the climatic favorability of tree dominated biomes to support the establishment and growth of trees. In many of the non-tree, dominated biomes irrigation is required for the establishment and maintenance of trees.

Nearly an equal number of the species found on waste areas had wind dispersed seed (51%) as had other mechanisms for seed dispersal (41%). Wind dispersal species were more common on waste sites in the deciduous forest, conifer forest, and mixed deciduous-conifer forest biomes than in other biomes (Table 5.13). Seventy percent of the waste area species in these three biomes were wind dispersed. The non-wind dispersed species were moved onto waste areas by birds or small mammals. Two somewhat unique modes of seed dispersal were observed in Bangkok and Dubai. Mangrove seedlings (*Rhizophora mucronata*) become established in abandoned canals in Bangkok by the floating of mangrove seeds into the shallow waters of these canals. In Dubai, boys use waste area as soccer fields. They often erect goalposts at the ends of these fields. The areas adjacent to these goalposts were observed to support Christ-thorn tree *(Ziziphus spina-christi)*. Boys who come to watch the soccer games bring the fruit of the Christ-thorn tree to eat. The discarded seeds germinate at the margins of the soccer fields adjacent to the goal posts (Table 5.14).

Table 5.13 Most common tree species in small parks

Biome	TF			BEF			DF			CF			MDCF			MWS			STS			G			D			T			H			
City	1	2	3	4	5	6	7	8	9	10	11	12	13	14	15	16	17	18	19	20	21	22	23	24	25	26	27	28	29	30	31	32	33	Sum
Aesculus hippocastanum							•	•				•		•	•																			5
Albizia julibrissima	•						•										•						•	•								•		6
Casuarina equisetifolia			•	•		•												•		•														5
Cedrus deodara					•							•					•					•	•	•							•			7
Delonix regia		•	•			•												•			•			•	•	•	•							9
Ficus elastica		•	•															•				•			•									5
Fraxinus pannsylvanica							•	•			•	•										•												5
Liquidambar styraciflua										•	•	•				•	•																	5
Magnolia grandiflora										•	•					•	•	•				•		•								•		8

Oreodoxa regia	•	•																	•	•			•		•									6
Pheonix canarienses										•						•	•	•	•			•	•			•								8
Platanus x hispanica				•			•	•	•			•				•	•					•	•											9
Populus nigra 'Italica'															•	•			•			•		•										5
Prunus serrulata									•			•	•		•								•								•			6
Robinia pseudoacacia					•		•		•			•				•	•					•	•	•	•							•		11
Salix babylonica	•					•						•										•		•							•			6
Washingtonia fiilfera																	•					•	•		•	•	•							6
Sum	3	3	3	2	2	3	5	3	3	3	3	8	1	1	3	6	8	5	3	2	1	10	8	7	5	3	2	0	0	0	3	3	0	

• means the species is present

1 = Bangkok, 2 =Rio, 3 = Singapore, 4 = Tokyo, 5 = Sydney, 6 = Hong Kong, 7 = London, 8 = New York, 9 = Paris, 10 = Jacksonville, 11 = Seattle, 12 = Vancouver, 13 = Beijing, 14 = Moscow, 15 = Stockholm, 16 = Athens, 17 = Los Angeles, 18 = Tel Aviv, 19 = Brasilia, 20 = New Delhi, 21 = Lagos, 22 = Buenos Aires, 23 = Johannesburg, 24 = Tehran, 25 = Cairo, 26 = Dubai, 27 = Phoenix, 28 = Murmansk, 29 = Nuuk, 30 = Reykjavik, 31 = Lhasa, 32 = Quito; 33 = Cusco

Table 5.14 Species observed on waste areas

Biome	TF			BEF			DF			CF			MDCF			MWS			SS			G			D			T			H			
City[a]	1	2	3	4	5	6	7	8	9	10	11	12	13	14	15	16	17	18	19	20	21	22	23	24	25	26	27	28	29	30	31	32	33	Σ
Abies lba															•																			1
Acacia confusa						•																												1
Acacia dealbata																																•		1
Acacia farnesiana																										•								1
Acacia visco		•																																1
Acer macrophyllum											•																							1
Acer negundo														•																				1
Acer platanoides														•	•																			2
Acer pseudoplatanus							•																											1
Acer saccharum var. florida										•																								1
Ailanthus altissima							•	•	•							•	•	•				•												7
Albizia julibrissin	•																			•														2
Aleurites moluccana						•																												1
Alnus japonica					•																													1
Alnus rubra											•																							1
Artocarpus altilis	•		•																															2
Avicennia marina	•																																	1
Azadirachta indica																				•														1
Betula pendula														•	•																			2
Betula pubescens																												•	•					2
Bombax ceiba						•																												1
Caria papaya	•																																	1
Cassia fistula	•																																	1

City[a]	1	2	3	4	5	6	7	8	9	10	11	12	13	14	15	16	17	18	19	20	21	22	23	24	25	26	27	28	29	30	31	32	33	Σ
Cassia surattensis																					•													1
Casuarina cunninghamiana																					•													1
Celtis sinensis					•																													1
Cericidium floridum																											•							1
Chukrasia tabularis			•																															1
Clerodendrum trichotomum					•																													1
Corylus avellana															•																			1
Delonix regia						•																												1
Eucalyptus globulus																																•	•	2
Fagraea crenulata			•																															1
Ficus benghalensis																				•						•								2
Ficus carica										•											•													2
Ficus pseudo-sycomorus																		•																1
Ficus religiosa																				•														1
Fraxinus excelsior							•								•																			2
Ilex integra					•																													1
Juglans neotropica																																•		1
Juglans regia							•																	•										2
Laurus nobilis																•																		1
Mallotus japonicus					•																													1

(continued)

Table 5.14 (continued)

Biome	TF			BEF			DF			CF			MDCF			MWS			SS			G			D			T			H			
City[a]	1	2	3	4	5	6	7	8	9	10	11	12	13	14	15	16	17	18	19	20	21	22	23	24	25	26	27	28	29	30	31	32	33	Σ
Musa paradisiana																					•													1
Nandina domestica					•																													1
Nyssa sylvatica										•																								1
Peltophroum pterocarpa						•																												1
Pinus sylvestris															•																			1
Pithecellobium dulce																										•								1
Pittosporum crassifolium																	•																	1
Pittosporum undulatum																	•																	1
Platanus occidentalis										•																								1
Populus deltoides								•		•																								2
Populus nigra 'Italica'									•																									1
Populus tomentosa													•																		•			2
Populus trichocarpa											•																							1
Populus trichocarpa																													•					1
Populus x *canadensis*									•																									1
Prosopis pubescens																											•							1
Prunus padus															•																			1
Quercus agrifolia																	•																	1
Quercus robur															•																			1
Quercus virginiana										•																								1
Robinia pseudoacacia									•				•									•												3
Sabal palmetto										•																								1

City[a]	1	2	3	4	5	6	7	8	9	10	11	12	13	14	15	16	17	18	19	20	21	22	23	24	25	26	27	28	29	30	31	32	33	Σ
Salix caprea																												•						1
Salix scouleriana											•																							1
Salix tetrasperma																				•														1
Sambucus nigra																																•		1
Schefflera actinophylla																					•													1
Sorbus aucuparia														•	•													•	•					4
Tamarix aphylla																		•																1
Tamarix articulata																											•							1
Ulmus pumila																															•			1
Washingtonia robusta																	•																	1
Ziziphus spina-christi																										•								1
City[a]	1	2	3	4	5	6	7	8	9	10	11	12	13	14	15	16	17	18	19	20	21	22	23	24	25	26	27	28	29	30	31	32	33	
Total for each city (Total Number of Species = 75)	5	1	3	0	6	5	4	2	4	7	4	0	2	4	9	2	5	3	0	4	5	2	0	1	0	4	3	3	3	0	1	4	1	

• means the species is present

[a]1 = Bangkok; 2 = Rio de Janeiro; 3 = Singapore; 4 = Sydney; 5 = Tokyo; 6 = Hong Kong; 7 = London; 8 = New York; 9 = Paris; 10 = Jacksonville; 11 = Seattle; 12 = Vancouver; 13 = Beijing; 14 = Moscow; 15 = Stockholm; 16 = Athens; 17 = Los Angeles; 18 = Tel Aviv; 19 = Brasilia; 20 = New Delhi; 21 = Lagos; 22 = Buenos Aires; 23 = Johannesburg; 24 = Tehran; 25 = Cairo; 26 = Dubai; 27 = Phoenix; 28 = Murmansk; 29 = Reykjavik; 30 = Nuuk; 31 = Lhasa; 32 = Quito; 33 = Cusco

Waste areas are used in some cities for gardening by people who have no other access to land. Waste area gardens were observed in Bangkok, New Delhi, and Quito. The tending of these gardens prevents the natural establishment of tree species in these waste areas. In Bangkok papaya, cassia, breadfruit, and bananas were observed in the waste area gardens. Small patches of corn were growing on waste areas in New Delhi and Quito. Many waste areas in New Delhi were being grazed by feral cattle. Some waste areas had been cut for fodder to feed livestock. These activities constrain the natural establishment of trees.

Nine tree species were found on waste areas in two or more cites. Ailanthus *(Ailanthus altissima)* and rowan (*Sorbus aucuparia)* were the most frequently encountered species occurring in six and four cities respectively. Ailanthus was found on waste areas in three biomes. Its widespread occurrence is due in part to its introduction to many cities in Europe and the western hemisphere in the later half of the eighteenth century. It was initially sought as a tree for gardens and streets. The initial encouragement of the planting of ailanthus by influential journals such as the *New England Farmer* (1833) in the United States was followed by an unjustified scorn of the species by the eminent American horticulturalist Andrew Jackson Downing (Downing 1852a, b). Jackson was swept up in a wave of American nationalism and disparaged the planting of ailanthus, along with other exotic species, in favor of the use of trees native to North America. In spite of Downing's prejudice against the species it gained a foothold in American cities prior to its fall from grace. During the California Gold Rush and the railroad-building era in western North America it was introduced over a wide area by Chinese laborers. Frederick Law Olmsted attempted to reverse the prejudice against the tree by using it in Central Park in New York City as well as in many of his other North American projects (Fabos et al. 1968; Fein et al. 1972). His attempts to revive interest in the species were not entirely successful because of the susceptibility of ailanthus to damage during ice storms, the unpleasant odor associated with male trees, and its invasive nature.

Its invasive characteristics, wind-dispersed seed, capacity to become established on harsh sites, tolerance of a wide range of climatic conditions, and competitive ability have made Ailanthus one of the most common colonizers of waste sites in cities. It also colonized railroad-right-of-ways, spreading from town to town in many regions. The ability of the species to sprout vigorously from stumps and roots has made it a difficult species to eradicate. Clearing of waste areas where ailanthus has become established is often followed by an army of ailanthus sprouts. Its invasive character has caused it to be spurned by many arborists and it is often listed as a weed species to be eliminated (Bossard et al. 2000). It has, however, gained a place in the hearts of Americans through its symbolism of hope in the novel *A Tree Grows in Brooklyn* (Smith 1943).

> It grew lushly, but only in the tenement districts…It came there first. Afterwards, poor foreigners seeped in and the quiet old brownstone houses were hacked up into flats, feather beds were pushed out on the window sills to air and the Tree of Heaven flourished. That was the kind of tree it was. It liked poor people.

Rowan, the second most widely spread waste area species encountered in this study, was found colonizing waste areas in four cites (Stockholm, Moscow, Murmansk, Reykjavik) across two biomes (mixed deciduous-conifer forest, tundra). It was in the top five species on the basis of frequency of occurrence in each of the four cities where it is also found colonizing waste areas. Rowan is a popular tree for parks and gardens in Stockholm and Moscow. In Murmansk and Reykjavik it is used both as a street trees as well as in gardens and parks. It is prized for its hardiness, well-proportioned foliage, red berries, and autumn color (Jacobson 1989). Unlike ailanthus, it is not a wind-dispersed species. Its seeds are commonly vectored by birds. Rowan shares with ailanthus an ability to become established on harsh sites exhibiting a variety of soil conditions (Aas and Riedmiller 1994). Its frost hardiness contributes to its success as a colonizer of waste areas in many cites in northern Europe.

The natural establishment of trees on waste sites in different cities around the world eventually lead to relatively stable vegetation types composed of mixtures of native and exotic species. Studies in various cities suggest the importance of the proximity of different tree species to the establishment of their offsprings in waste-area (McBride and Gerhard 1995; Bradshaw 1976; Sukopp 1973; Sukoop 2002). Few urban waste area sites are left to succeed to a late seral stage vegetation because of fire hazard or the value of real estate. In some cases, urban waste areas have become city parks and the naturally established tree cover incorporated into the design of the park (Hough 1995). Urban waste areas have a high potential for environmental education and have been the focus of such programs in several cities (Vessel and Wong 1987). They also serve as valuable play areas for children (Louv 2005). The trees common to waste area provide materials for impromptu crafts as well as other imaginative uses by children.

Old agricultural drainage canals can still be observed in some parts of Bangkok and can be considered a special category of waste areas. These were originally dug and maintained by farmers to drain orchards and other agricultural land during the monsoon season. They also served as waterways to access the agricultural landscape. When maintained, they were generally about two meters wide and one meter deep at the low water stage. As urbanization has spread over agricultural land adjacent to Bangkok many of these drainage canals have been filled in. Most of the drainage canals that were not filled for the building of streets or structures have undergone hydric succession as a result of siltation and the invasion of trees. Typical species observed in the various stages of hydric succession in abandoned drainage canals are listed in Table 5.15. The extent to which the various seral stages in the development of successional forests will remain as a part of the Bangkok urban forest is difficult to predict. The city's Public Parks department does not currently have plan for the management of these areas. At some locations local residents have cleared the successional forest species from the filled in drainage canals and planted bananas and other fruit trees to provide for their household needs.

Table 5.15 Number of species, percent native species, and percent wind disseminated species found on waste areas

Biome	City	Number of species	Percent native[a]	Percent wind disseminated
Tropical forest	Bangkok	5	20	0
	Rio de Janeiro	1	100	0
	Singapore	3	67	67
	Average	**3**	**62**	**22**
Broadleaf evergreen forest	Hong Kong	5	20	20
	Sydney	–	–	–
	Tokyo	5	80	20
	Average	**5**	**50**	**20**
Deciduous forest	London	4	75	75
	New York	2	50	100
	Paris	4	0	75
	Average	**3.5**	**42**	**83**
Coniferous forest	Jacksonville	7	71	43
	Seattle	4	100	100
	Vancouver	–	–	–
	Average	**5.5**	**86**	**72**
Mixed deciduous coniferous forest	Beijing	2	50	50
	Moscow	5	60	80
	Stockholm	9	88	77
	Average	**5.3**	**66**	**60**
Mediterranean woodland scrub	Athens	2	50	50
	Los Angeles	5	20	20
	Tel Aviv	3	0	33
	Average	**3.3**	**23**	**34**
Savanna-thorn scrub	Brasilia	–	–	–
	Lagos	5	0	20
	New Delhi	3	67	33
	Average	**4**	**34**	**26**
Grassland	Buenos Aires	2	0	50
	Johannesburg	–	–	–
	Tehran	1	0	0
	Average	**1.5**	**0**	**25**
Desert	Cairo	0	0	0
	Dubai	4	75	0
	Phoenix	3	71	0
	Average	**2.3**	**88**	**0**
Tundra	Murmansk	3	0	67
	Nuuk	0	–	–
	Reykjavik	3	0	67
	Average	**3**	**0**	**67**

(continued)

Table 5.15 (continued)

Biome	City	Number of species	Percent native[a]	Percent wind disseminated
Highlands	Cusco	1	0	100
	Lhasa	3	33	100
	Quito	4	50	25
	Average	**3.5**	**42**	**75**
Overall average		**3.6**	**44.8**	**44**

[a]Native to biome; – no samples

Water's Edge

Many of the world's cities were built on rivers, lakes, and estuaries. One cannot think of Paris without thinking of the Siene, Seattle without Lake Washington and Elliot Bay, or Sydney without Sydney Harbor. The juxtaposition of water with the terrestrial landscape, be it rural or urban, provides a scene that is highly ranked in studies of landscape preference and remain in our memory (Litton and Tetlow 1978). Trees along the margin between the water and the land add to that high preference. Cities have also been built on lesser known rivers and streams for which there is little association except in the minds of the local people. These often become valuable environments for children who seek them out as special places for play. The poet, Anne Michaels (1997), captured the symbolic importance of these unnamed urban streams when she wrote:

> These weekly explorations into the ravines were escapes to ideal landscapes; lakes and primeval forests so long gone they could never be taken away from us.

The margins of bodies of water in urban areas were valued sites for the loading and unloading of ships and often were utilized for industrial purposes in many cities. Smaller streams in urban areas were often covered over in order to control flooding and provide space for building (Barton 1962). The nature of transportation and industry changed in the middle half of the twentieth century resulting in the freeing of land for other uses along many rivers, lakes, and estuaries. Waterfront areas saw the development of new commercial centers, parks, and promenades where trees were planted to enhance the waterfront environment. The environmental movement of the 1970s stimulated the cleaning of forgotten streams in urban areas and many projects to uncover buried streams and restore urban riparian vegetation (Riley 1998).

The objective of surveying sites along the water was to characterize the use of trees in this special environment and to contrast these characteristics with those of trees found in other urban environments. The characteristics examined include species composition, percentage of riparian species used, pattern of planting, spacing, and extent of canopy cover. The design aspect of water's edge environments will be addressed in a subsequent chapter, the emphasis here will be on species composition.

Table 5.16 Typical species observed in the various stages of hydric succession in abandoned drainage canals in Bangkok

Scientific name	Common name
Wetter zones:	
Alternanthera sessilis	Sessile joywood
Arundo donax	Giant reed
Colocasia antiquorum	Colocasia
Cyperus piloisus	Hairy sedge
Cyperus procerus	Lax sedge
Eclipta prostrata	False daisy, white eclipta
Ipomoea aquatica	Wooly morning glory
Ludwigia adscendens	Creeping water primrose
Typha angustifolia	Narrow-leaved cat tail
Drier zones:	
Brachiaria mutica	Water grass, buffalo grass, para grass
Coix aquatica	Water coix
Imperata cylindrica	Cogongrass
Trees:	
Nypa frutican	Nipa
Sonneratia caseolaris	Sonneratia

One hundred and twenty-one species were observed in water's edge environments in this study. Of these 37% are riparian species. This compares to a representation by riparian species of 20% of the total number of species observed in all urban environments. The larger percentage of riparian species found in water's edge environments is due to the natural occurrence of riparian species along streams in the suburban portions of cities and a conscious decision on the part of urban forest planners to plant riparian species. Water's edge sites in several cities (e.g., Moscow, Phoenix, Beijing, Jacksonville, Quito, Stockholm) included sites with natural stream banks that allow for the continued existence of native riparian species. Some cities have emphasized the water's edge environment by planting riparian species. Beijing made extensive use of weeping willow in this context. In Paris, along the Seine, Carolina poplar is widely used.

Four species were observed in water's edge environments in three or more cities (Table 5.16). Three of these, weeping willow, casuarina, and London plane, are riparian species. None of the four species were recorded in natural woodlands in the cities studied. Their common occurrence was the result of design decisions. Weeping willow, recorded in water's edge environments in five cities and in five biomes, was often used exclusively in these cities to delineate rivers, stream, and the margins of estuaries and lakes. Casuarina, London plane, and royal poinciana were used in this manner, but were often found in other urban environments as well.

A unique aspect of the water's edge environment was its higher level of soil moisture and potential for flooding. The characteristics of sites adjacent to natural streams dictate the species composition of native riparian woodlands (Conard et al. 1980). The importance of these characteristics in urban areas is often diminished by flood control dams and the more impervious materials used to harden the stream banks in some cities. The higher soil moisture conditions of unarmored streams banks in suburban sections of many cities contribute to the dominance of riparian woodland species along these streams. The frequency of flooding in Bangkok has limited the palette of species that can be used along the water's edge even in the most built up portion of the city in spite of the channelization of the streams and canals. Rivers and stream in cities built on estuaries or the ocean may experience saltwater migration up stream during periods of high tide depending on the volume and rate of fresh water flow. Saltwater intrusion into the water table adjacent to these streams can severely limit the tree species that can be planted along these streams. In Dubai, saltwater intrusion has prevented the common use of ghaf *(Prosopis chineraria)*, neem *(Azadirachta indica)*, and toothbrush tree (*Salvadora persica*) along major highways through parts if the city. These species have been replaced by tamarix *(Tamarix aphylla)* and casuarina *(Casuarina cunninghamiana)* in areas of saltwater intrusion.

A few water's edge sites in the survey were located on natural and semi-natural portions of streams. These supported native species, which in some cases had been invaded by exotic species. Ailanthus, box elder, black locust, and Italian poplar were common exotic species observed invading these sites in temperate zone cities such as Moscow, Beijing, and Seattle.

Cemeteries

Urban cemeteries in much of Europe and the United States support rich mixtures of trees that often are the most diverse of any environment within a city. Stately trees, together with lakes and architectural features drew American city dwellers to cemeteries for strolling as a form of recreation by the middle of the nineteenth century. Historians of public landscapes credit the popularity of these cemeteries with creating a demand for public parks and giving birth to the profession of landscape architecture in the United States (Newton 1971; Linden-Ward 1989). Rural cemeteries built on the outskirts of Paris (Pere Lachaise), London (Highgate), and Boston (Mount Auburn) became models of tree species richness that were emulated in twentieth century. The continued importance of cemeteries to the diversity of urban tree flora and fauna has been emphasized by Jackson and Vergara (1989) and Spirn (1984). Jacobson (1989) praises cemeteries as, "good places for trees and tree enthusiasts". He points out that most cemeteries are intended to be permanent and cemetery trees are provided with ample growing space for full development of their canopies.

The cemeteries surveyed in this study reflect the multiple interactions of religious traditions, historical development of cities, and the physical and biological constraints of different biomes. The sample of 33 cities did not provide enough information to clearly understand the interactions of these factors. Few clear patterns of the use of trees in cemeteries of the world's cities are apparent in the data. A total of 172 species of trees were recorded in the cemeteries surveyed. Species richness varied from a low of two species observed in a cemetery in Cairo's "City of the Dead" to a high of 24 species in London's Highgate Cemetery.

The number of species occurring in cemeteries is dependent upon the management of the cemeteries, the climatic restrictions of the biome, and the planting space available for trees. Cemeteries in Tehran, for example, are under the management of the municipal government. The city administrators makes decisions on what trees are planted, the layout of the plantings, and the maintenance of the cemeteries. Only four tree species were being used in Tehran. These were laid out along grid lines to define units within the cemeteries containing 144 graves. In contrast, the cemeteries studied in London, Moscow, New York, and Johannesburg allowed family members to plant trees on family grave plots. These four cities had the highest cemetery tree species richness in the survey (London = 24, New York = 22, Johannesburg = 19, Moscow = 17). The average number of tree species observed in cemeteries in the 28 cities studied was 9.8. Management of gravesites and cemeteries in general can influence tree species richness. Many older cemeteries lack current maintenance programs for tree management. This responsibility was previously delegated to surviving family members who themselves may have passed away. Some of the older cemeteries in London, New York, Jacksonville, Singapore, and Bangkok showed evidence of the lack of maintenance and the voluntary establishment of trees. Portions of the Highgate Cemetery in London could be characterized as thickets of European ash supporting dense understories of English ivy growing over gravestones to create nearly impenetrable portions of the cemetery. Voluntary trees were in some cases tipping over tombstones. Ailanthus, the most common tree species observed on urban waste areas, also showed up as a volunteer species in cemeteries in three of the cities studied. The invasion of cemeteries by trees in cities in tree dominated biomes is to be expected where cemeteries are not maintained. This was evident for cemeteries surveyed in cities in the tropical forest, deciduous forest, conifer forest, and mixed deciduous-conifer forest biomes.

Climatic restrictions played a role in the species richness of trees in cemeteries in Murmansk, Nuuk, and Reykjavik. Cemeteries in these cities supported no trees or at most only five species. Cemetery management policies that allowed family members to plant trees of their choice on individual grave plots, but climatic restrictions limited the success of tree planting, especially in Nuuk. The low number of tree species observed in Cairo's "City of the Dead", only two species, was a reflection of the lack of irrigation water to support the establishment and survival of trees in this desert environment. Cemeteries in Dubai and Phoenix were equipped with irrigation systems and supported larges numbers of species (Dubai = 9, Phoenix = 11). The space available for tree planting also is a factor that appears to have influenced tree species richness in some cemeteries.

Twenty-one of the 172 species recorded in cemeteries occurred in cemeteries in three or more of the cities studied (Table 5.17). Italian cypress, the most commonly encountered cemetery tree, was found in cemeteries in 9 of the 28 cities. These cities were spread across six of the world's biomes (deciduous forest, Mediterranean woodland scrub, savanna thorn scrub, grassland, desert, highland). Italian cypress

Table 5.17 Species occurring in water's edge environments in three or more cities

Species		Cities	Biomes
Common name	Scientific name		
Weeping Willow	*Salix babylonica*	5	5
Casuarina	*Casuarina equisetifolia*	5	4
London Plane	*Platanus x hispanica*	3	2
Royal Poinciana	*Delonix regia*	3	1

has long been associated with cemeteries in Europe. It was often planted in ancient Greece as a memorial to a deceased hero or relative (Dwyer et al. 1994). Its popularity as a tree for cemeteries in Europe and North America is based on its dense, dark-evergreen foliage and its long life. Its somber appearance symbolizes mourning, despair, death, and eternal sorrow. Many people see in its evergreen foliage a symbol of everlasting life. Dwyer et al. (1994) in their investigation of the spiritual importance of cemetery trees, such as Italian cypress, found trees were very important in the process of making psychological adjustment to the death of family members and friends.

Canary Island date palm occurred in cemeteries in eight of the cities in the study. These cities occurred in five biomes (broadleaved evergreen forest, Mediterranean woodland scrub, grassland, desert, undifferentiated highlands). It is a stately tree that is often used to line roadways through cemeteries. Its association with burial dates back to the ancient Egyptians who wove strips of the leaves of date palms into broad floral collarets to be placed around the effigy masks of the dead (Hepper 1990).

Twenty percent of the 172 species observed in cemeteries were conifer species. This compares to only 12.6% of the tree species observed in all of the urban environments in the study. The use of conifer species in cemeteries symbolizes everlasting life and adds to the somberness of the cemetery landscape. The percentage of palm species used in cemeteries (4.7%) was nearly double their use in all urban environments (2.5%). Their use may be due to tradition in the Middle East. The common use of palms may also stem from the conformity of their root systems to small spaces, absence of lower branches that might interfere with vehicles and pedestrians, and the ease of transplanting large specimens.

Freeways

Limited access, multi-lane highways cross and skirt around the world's cities. They are called "freeways", "motorways", "turnpikes", "toll roads", "ring roads" and "parkways" in English-speaking countries. The terms "freeway" will be used here to as a general term to denote limited access, multi-lane highways in urban areas. The margins of urban freeways are often planted with trees that make a significant contribution to the urban forests.

Tree-lined parkways were designed in the first half of twentieth century as automobile touring became a national pastime in the United States. Although most early parkways were in rural locations (e.g., Blue Ridge Parkway, Virginia; Garden State Parkway, New Jersey; Natchez Trace Parkway, Mississippi) a few were located in urban and suburban areas (e.g., Bronx River Parkway, New York; Arroyo-Seco Parkway, Los Angeles) (Jolley 1989). Parkway design focused on maintaining and enhancing the scenic values along a road. Most early parkways had traffic lanes divided by planted median strips, somewhat limited access, and cross roads passing over or under the parkway (Fig. 5.1). Several parkways constructed during the American Depression remain outstanding highways that attract large numbers of motorists because of their scenic beauty.

Trees used along freeways are selected because of their minimum maintenance requirements, minimum potential for wind breakage and wind throw, lack of large fruits or cones that would litter the roadway, branching structures that do not interfere with visibility, attractive-showy flowers, and minimum water requirement (in drier biomes). Species observed along freeways in Los Angeles are noted for some degree of drought tolerance (Perry 1989). Summer irrigation was required to establish these trees, once established they are not irrigated. However, some portions of freeway right-of-ways are irrigated to support ground cover.

To sample trees along freeways in Los Angeles one mile long sections of right-of-ways were used. Trees occurring along every fifth mile of right-of-way were recorded. Some one-mile sections did not support any trees while the maximum number recorded for any one-mile section was 10. Seventy-one tree species were observed growing along the right-of-ways of six freeways in the Los Angeles metropolitan area (Table 5.18). The average number of trees recorded was six. The

Fig. 5.1 Arroyo Seco Parkway (c. 1940) (Redrawn from postcard)

Table 5.18 Number of cities and biomes in which the most common cemetery trees occurred

Species		Number of	
Common name	Scientific name	Cities	Biomes
Italian Cypress	*Cupressus sempervirens*	9	6
Canary Island Date Palm	*Phoenix canariensis*	9	6
Norway Maple	*Acer platanoides*	6	4
Crape myrtle	*Lagerstroemia indica*	4	4
Italian Stone Pine	*Pinus pinea*	4	3
Jacaranda	*Jacaranda mimosifolia*	4	4
London Plane	*Platanus* x *hispanica*	4	3
Rowan	*Sorbus aucuparia*	4	3
Ailanthus	*Ailanthus altissima*	3	2
Black Locust	*Robinia pseudoacacia*	3	3
Blue Gum	*Eucalyptus globulus*	3	3
Casuarina	*Casuarina cunninghamiana*	4	3
Deodar Cedar	*Cedrus deodara*	4	3
European Ash	*Fraxinus excelsior*	3	3
Field Maple	*Acer campestre*	3	3
Hawthorn	*Crataegus monogyna*	3	3
Horse Chestnut	*Aesculus carnea*	3	2
Little Leaf Linden	*Tilia cordata*	4	3
Monterey Pine	*Pinus radiata*	3	3
Scots Pine	*Pinus sylvatica*	3	3
Washington Palm	*Washingtonia filifera*	5	5
Hollywood Juniper	*Juniperus chinesis* 'Torulosa'	3	3

percentage frequencies of occurrence of the top five species along the six freeways sampled were lemon scented gum (51), Mexican fan palm (49), red gum (42), California pepper tree (34), and Chinese elm (32) (Table 5.19).

Only four species were observed along the freeways surveyed in Beijing. These were Chinese poplar, Chinese arborvitae, Japanese pagoda tree, and weeping willow. Sections of the right-of-ways along Beijing's Fifth Ring Road have been planted as a part of the greenbelt to define the limit of urban growth. In these sections the freeway is flanked by five rows of Chinese poplar trees. A center divider, which was planted with Chinese arborvitae, separated the opposing lanes of traffic. Trees planted in this fashion, especially the flanking rows of Chinese poplar, help reduce the noise of traffic and to control dust from the freeway. An earlier planting design, that accommodates bicycle traffic, can be seen along sections of Beijing's Fourth Ring Road. Here single rows of Japanese pagoda trees and weeping willows have been planted adjacent to bicycle lanes along each side of the freeway. The Chinese arborvitae was also planted in the center divider of the Fourth Ring Road.

Twenty-two tree species occurred along the freeways sampled in Singapore. Trees used along freeway right-of-ways in Singapore are often planted to provide screening for adjacent housing and recreational areas. Butterfly pea tree (*Clitoria racemosa*), eugenia (*Eugenia oleina*), and ordeal tree (*Erythrophleum guineense*)

Table 5.19 Trees observed along six Los Angeles Freeways

Species		Freeway					
Common name	Scientific name	1	2	3	4	5	6
Allepo pine	*Pinus halepensis*	•		•	•		•
Australian willow	*Geijera parviflora*			•			•
Black locust	*Robinia pseudoacacia*	•		•	•		•
Bailey acacia	*Acacia baileyana*	•		•	•		
Blackwood acacia	*Acacia melanoxylon*	•			•		•
Bo tree	*Ficus religiosa*	•		•	•		•
Bushy yate	*Eucalyptus lehmannii*	•			•		
California pepper tree	*Schinus molle*	•	•	•	•	•	•
California sycamore	*Platanus racemosa*	•		•	•		•
California walnut	*Juglans californica*	•	•	•	•	•	•
Canary Island Date palm	*Phoenix canariensis*	•	•	•	•	•	•
Canary Island pine	*Pinus canariensis*	•	•	•	•	•	•
Catalina Island Ironwood	*Lyonothamnus floribundus*					•	
Chinese elm	*Ulmus parvifolia*	•	•	•	•	•	•
Chinese flame tree	*Koelreuteria bipinnata*	•			•		
Coast live oak	*Quercus agrifolia*	•					
Carob	*Ceratonia siliqua*	•	•	•	•	•	•
Crape myrtle	*Lagerstroemia indica*	•		•	•		•
Crimson flowered eucalyptus	*Eucalyptus leucoxylon*			•			•
Crown of Gold	*Cassia excelsa*			•			•
Elderberry	*Sambucus nigra*						•
Empress tree	*Paulownia tomentosa*	•	•	•	•	•	•
Tree ceanothus	*Ceanothus thrysiflorus*	•					
Golden medallion tree	*Cassia leptophylla*	•			•		
Guadalupe palm	*Brahea edulis*	•			•		
Holm oak	*Quercus ilex*			•			•
Honey Mesquite	*Prosopis glandulosa*	•			•		
Hong Kong Orchid Tree	*Bauhinia x blakeana*			•			•
Horsetail tree	*Casuarina equisetifolia*	•			•		
Italian stone pine	*Pinus pinea*	•			•		
Jacaranda	*Jacaranda mimosifolia*						•
Japanese black pine	*Pinus thunbergii*			•			•
Jelecote pine	*Pinus patula*	•		•	•		•
Karo	*Pittosporum crassifolium*	•		•	•		•
Lemon bottle brush	*Callistemon citrinus*	•			•		
Lemon-scented gum	*Corymbia citriodora*	•	•	•	•	•	•
Manna gum	*Eucalyptus viminalis*	•			•		
Monterey pine	*Pinus radiata*	•	•		•	•	
Mexican fan palm	*Washingtonia robusta*	•	•	•	•	•	•

(continued)

Table 5.19 (continued)

Species		Freeway					
Morten bay chestnut	*Ficus macrophylla*	•		•	•		•
Oleander	*Nerium oleander*	•			•		
Olive	*Olea europaea*	•		•	•		•
Paperbark tree	*Melaleuca quinquenervia*	•			•		
Purple Bailey acacia	*Acacia baileyana 'Purpurea'*			•			•
Red-flowering Gum	*Corymbia ficifolia*			•			•
Royal palm	*Roystonea regia*	•			•		
Royal poinciana	*Delonix regia*	•			•		
Red gum	*Eucalyptus camaldulensis*	•	•	•	•	•	•
Red ironbark	*Eucalyptus sideroxylon*	•	•	•	•	•	•
Redwood	*Sequoia sempervirens*	•			•		
Shamel ash	*Fraxinus uhdei*			•			•
She oak	*Casuarina equisetifolia*	•	•	•	•	•	•
Smallflower tamarisk	*Tamarix parviflora*				•		
Snow in summer	*Cerastium tomentosum*					•	
Silk floss tree	*Ceiba speciosa*			•			
Silk oak	*Grevillea robusta*			•			•
Silk tree	*Albizia julibrissin*			•			•
Silver dollar tree	*Eucalyptus cinerea*	•	•	•	•	•	•
Silver dollar gum	*Eucalyptus polyanthemos*	•			•		
Strawberry tree	*Arbutus unedo*						•
Sweetgum	*Liquidambar styraciflua*	•			•		
Tobacco tree	*Nicotiana glauca*				•		
Victorian-box	*Pittosporum undulatum*		•				
Washington palm	*Washingtonia filifera*	•			•		
White alder	*Alnus rhombifolia*			•			•
White ironbark	*Eucalyptus leucoxylon*	•					
White floss-silk Tree	*Ceiba insignis*		•				
Willow acacia	*Acacia salicina*						•
Windmill palm	*Trachycarpus fortunei*			•			•
Woman's tongue	*Albizia lebbeck*			•			•

• means the species is present
1=San Diego Freeway; 2 = Santa Monica-San Bernardino; 3 = Ventura; 4 = San Gabriel River; 5 = Harbor; 6 = Artesia

have been commonly planted for screening. The rain tree (*Samanea saman*) has been planted in the center divider of the East Coast Parkway as well as along sections of the margins of this freeway. Other trees commonly used for freeway planting in Singapore include common pulai (*Alstonia angustiloba*), India pulai (*Alstonia scholaris)*, sentang (*Azadirachta excelsa*), Brazilian ironwood (*Caesalpinia ferrea*), golden shower (*Cassia fistula*), wild cinnamon (*Cinnamomum iners*), fiddle-wood (*Citharexylum spinosum*), pink mempat (*Cratoxylum formosum*), salam (*Eugenis polyantha*), cabbage tree (*Fagraea crenulata*), tembusu (*Fragraea fragrans*), chengal pasir (*Hopea odorata*), senegal khaya (*Khaya senegalensis*), Ceylon ironwood (*Mesua ferrea*), carpentaria palm (*Carpentaria acuminata*), Chinese fan palm (*Livistona chinensis*), Fiji fan palm (*Pritchardia pacifica*), and cabbage palm (*Roystonea oleracea*).

Rooftop Gardens

Growing trees on the roofs of buildings dates back at least to the Hanging Gardens of Babylon built during the reign of Nebuchadnezzar II (605–562 BC) (Clifford 1963). The ancient Syrians and Greeks often adorned the roofs of their houses with statues and potted plants to honor the fertility of their gods, while wealthy Romans frequently constructed pleasure gardens atop their homes in urban areas. When these roof gardens included trees they were called '*Sylvae in tectis*' (roof forests) (Wright 1939). Hanging gardens were described by Diaz (1632) in Chapultepec, the summer residence of the Aztec King Montezuma. Diaz observed that flowers were grown on rooftops in other Aztec cities at the time of the conquest of Mexico by Cortez. A renewed interest in roof plantings occurred in Europe in the eighteenth century, but it was not until the twentieth century that roof top gardening really blossomed. As buildings began to occupy much of the downtown areas of cities like New York City and Los Angeles the use of rooftops gardens became a popular undertaking. Theaters, department stores, restaurants, hotels, and penthouses of the well-to-do utilized roofs for gardens (Haus-Rucker Inc., 1976). Following the Second World War, rooftop gardening grew in popularity and was adopted by apartment buildings, civic buildings, and parking structures. It is estimated that there are about 1500 private rooftop gardens on Manhattan Island in New York City (Du Val 2003).

At the end of the twentieth century an interest in the energy saving and urban heat island reduction potential of rooftop gardens resulted in the creation of plantings designs for rooftop to achieve these objectives. These gardens are sometimes referred to as "extensive green roofs" and are not designed for recreational use (Scholz-Barth 2001). These green roof designs do not generally include trees, but are reminiscent of sod roofs employed by the Viking settlers of Iceland and the pioneers who built sod houses on the North American prairies. A "green roof" movement started in Europe in the 1960s and spread to many cities in other parts of the world. By 1996 over 3.2 million ft^2 of green roofs had been constructed in Germany. It has been estimated that one in ten flat roofs in Germany are now vegetated (Port Philip Ecocentre 1998). Many Swiss cities now require new construction projects to relocate the area of green space covered by buildings to the rooftops. In 2000 the Japanese government required all new public and private commercial buildings with roof areas over 250 m^2 and 1000 m^2, respectively, to "green" a minimum of 20% of the area of any flat roof. The government of Hong Kong also issued similar standards for green roofs in 2000.

Table 5.20 Frequencies of occurrence of top five freeway species

Tree species		
Common name	Scientific name	Frequency (%)
Lemon-scented gum	*Eucalyptus citriodora*	51.2
Mexican fan Palm	*Washingtonia robusta*	48.8
Red Gum	*Eucalyptus camaldulensis*	41.5
California pepper tree	*Schinus molle*	34.1
Chinese elm	*Ulmus parvifolia*	31.7

Rooftop gardens usually experience colder temperatures and higher wind velocities than gardens planted on the surface of the ground in urban areas. Trees growing on rooftops are likely to experience water stress because of the limited soil moisture storage capacity of their containers and high evaporative rates associated with higher wind velocities. Proper tree species selection and frequent pruning are essential to the survival of trees in rooftop gardens. Japanese black pine is an example of a species that is well suited to the rooftop environments in temperate zone cities. This species is native to the southeastern coast of Japan where it gained its adaptation for survival in environments characterized by high wind velocities and soils with low soil moisture storage capacity and low fertility.

Rooftop gardens were observed in New York, Singapore, and Tokyo in the course of surveying cities for this book. Tree species commonly used for rooftop gardens in these cities are listed in Table 5.20. No species were common to all three cities; however, eight species (Japanese maple, silver birch, Hinoki false cypress, Kousa dogwood, Chinese juniper, Japanese black pine, Japanese flowering cherry, Japanese stewartia) were found on rooftop gardens in both New York and Tokyo. The overlap in the use of species in New York and Tokyo is in part a function of similar climatic conditions of the two cities. The colder climates of New York and Tokyo prevent the use of species observed in Singapore. Trees planted in rooftop gardens contribute to energy saving in the heating and air conditioning of buildings. A study in Singapore (Hien et al. 2002) showed significant reductions in temperature, heat flux, and annual energy reduction beneath trees in rooftop gardens (Table 5.21) Rooftop trees also act to reduce runoff and urban air pollution. Social and recreational benefits have been shown to accrue from having rooftop gardens with trees (Yuen and Hong 2002). It can be anticipated that rooftops will continue to be an urban environments in which trees will be utilized (Table 5.22).

Table 5.21 Trees commonly used for rooftop gardens in New York, Singapore, and Tokyo

Species				
Scientific name	Common name	Singapore	New York	Tokyo
Abies veitchii	Japanese fir			•
Acer palmatum	Japanese Maple		•	•
Baeckia frutescens	False Ru	•		
Bauhinia blakeana	Hong Kong Bauhinia	•		
Betula maximowicziana	Monarch Birch			•
Betula nigra	River Birch		•	
Betula pendula	Silver Birch		•	•
Callistemon citrinus	Bottlebrush Tree	•		
Carpinus betulus 'Fastiagiata'	European Hornbeam		•	
Cassia fistula	Golden Shower	•		
Cassia spectabilis	Popcorn Tree	•		
Cercidophyllum japonica	Katsuratree			•
Chamaecypris obtusa	Hinoki False Cypress		•	•
Citharexylum spinosum	Fiddle Wood	•		
Cordia sebestena	Geiger Tree	•		
Cornus florida	Flowering Dogwood		•	
Cornus kousa	Japanese Dogwood		•	•
Cratoxylon cochinchinense	Kemutong	•		
Cratoxylon formosum	Pink Mempat	•		
Cynometra cauliflora	Nam Nam	•		
Eleagnus umbellata	Autumn Olive		•	
Erythrina crist-galli	Cockscomb Tree	•		
Erythrina glauca	Coral Tree	•		
Eugenia oleina	Eugenia	•		
Ginkgo biloba	Ginkgo			•
Gustavia superba	Membrillo	•		
Juniperus chinesis	Chinese Juniper		•	•
Koelreuteria paniculata	Golden Rain tree		•	
Lagerstroemia speciosa	Rose of India	•		
Malus hybrids	Crabapple		•	
Pinus thunbergii	Japanese Black Pine		•	•
Pisonia alba	Lettuce Tree	•		
Plumeria rubra	Frangipani	•		
Populus tremuloides	Quaking Aspen		•	

(continued)

Table 5.21 (continued)

Species				
Scientific name	Common name	Singapore	New York	Tokyo
Prunus serrulata	Japanese Flowering Cherry		•	•
Prunus subhirtella 'Pendula'	Weeping Cherry		•	
Prunus x yedoensis	Yoshino Cherry			•
Pyrus calleryana	Callery Pear			•
Salix babylonica	Weeping Willow	•		•
Stewartia pseudocamellis	Japanese Stewartia		•	•
Tabebuia chrysotricha	Trumpet Tree	•		
Tabebuia pallida	Pink Trumpet Tree	•		
Thuja standishii	Japanese arborvitae			•
Xanthostemon chrysanthus	Golden Penda	•		

• means the species is present

Table 5.22 Reduction in temperature and heat flux beneath trees in roof top gardens in Singapore

Thermal property	Maximum reduction
Surface temperature of roof	31.0 °C
Ambient Temperature at 300 mm height	4.2 °C
Solar Radiation at 300 mm height	124.6 W/m^2
Heat Flux transferred through surface	15.4 W/m^2
Total Heat gain over a day	466.3KJ/M^2
Average energy consumption	14.5%

Hien et al. (2002)

Chapter 6
Design

Introduction

Trees have been used as design objects in urban environments since the establishment of cities. Their use as aesthetic objects has ranged from the individual specimen tree that resemble sculpture to large numbers of trees unifying and enhancing the visual properties of streets and boulevards. Countless outdoor spaces in cities are defined by trees. As objects of design they provide variation in size, shape, color, and texture. The use of trees in urban design has been documented by Arnold (1993). He describes the design qualities of trees along avenues, streets, and allees, in parks, and in public and quasi-public spaces other than parks (e.g., parking lots, traffic islands, medians). Arnold contrasts eight geometric patterns grouping trees in small parks using examples from the United States and Europe (Fig. 6.1). He also reviews the design element of trees used in larger urban parks using Central Park in New York City as an example. Jacobs et al. (2002) provides an account of how trees have been used as design elements along boulevards in cities around the world. These and other works (Gruffydd 1987; Littlewood 1988; Tsuru 1992; Zion 1995; Trowbridge and Bassuk 2004) treating the use of trees in urban design should be consulted for details of tree species selection, spacing, and planting patterns in common use.

A few examples of the use of trees in urban design that were encountered in the cities surveyed for this book will be presented here. They represent the creative imagination of famous landscape architects and urban planners as well as the unsung designs of landscape architects, engineers, arborists and urban foresters creating interesting and functional uses of trees in urban areas. It should be recognized that tree species selection is an important component of design. Species composition observed in different urban environments has been described in the preceding chapter. Separation of composition from design will seem arbitrary to many landscape architects and urban planners. The separation in this book was used to more economically present the data on species composition from an ecological standpoint.

J.R. McBride, *The World's Urban Forests*, Future City 8,
DOI 10.1007/978-3-319-52108-4_6

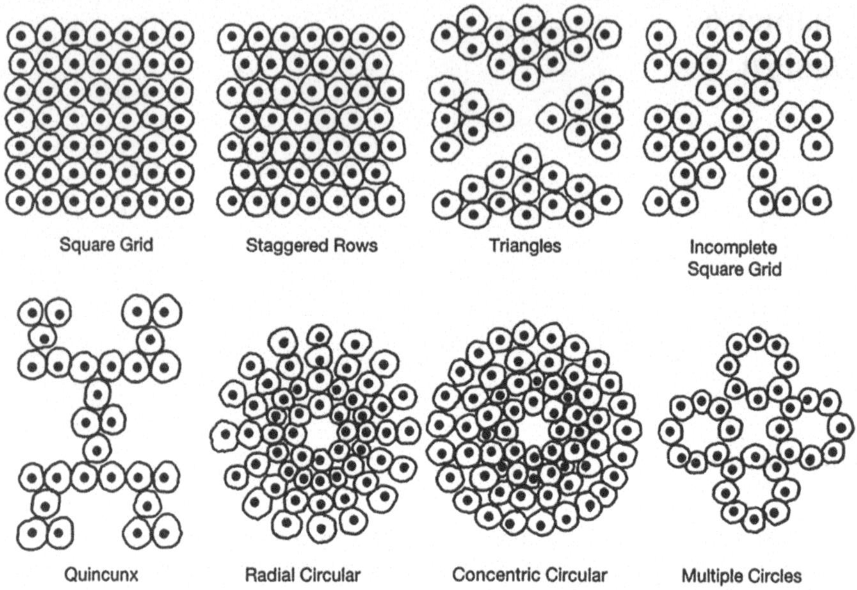

Fig. 6.1 Eight geometric patterns for grouping trees in small parks (Redrawn from Arnold 1993)

Some reference to tree species will be repeated in this chapter as it is deemed important to a better understanding of urban forest design.

Trees Along Streets and Boulevards

Trees are traditionally planted along streets in single or double rows. In cities like Beijing where the shading of sidewalks is important during the hot summers, triple rows of trees are used on wide sidewalks. Variations on typical designs for street tree planting were observed in New Delhi, Seattle, Stockholm, Murmansk, and Tehran. Before discussing these variations it will be useful to summarize the data collected on tree spacing and planting spaces used in different cities for planting street and boulevard trees.

Tree spacing and the dimensions of planting spaces used in different cities for planting street and boulevard trees are reported in Table 6.1. These data indicates that wider tree spacing is generally used in the planting of the medians of boulevards. Medians are usually wider than strips for street tree planting or than the widths of tree wells when planting strips are not used for street trees. Greater spacing of trees in boulevard medians stems from an interest in using larger trees in medians for both aesthetic and practical purposes, such as the shading of the boulevards. Wider space is needed for these trees to provide for adequate development of their large crowns. Wider spacing is also promoted by some traffic engineers as a

Table 6.1 Modal street and boulevard tree spacing and tree well dimensions

City	Tree spacing (ft.)		Tree well (ft.)	
	Blvd.	Street	Blvd.	Street
Singapore	40[1] (10 to 100)[2]	30 (15 to 80)	15∞	6∞ & 8 × 8 (3∞ to 25∞)
Bangkok	25 (20 to 25)	20 (12 to 40)	6∞	4 ×6 & 4∞ (2 × 2 to 6 × 6) (2∞ to 10∞)
Rio de Janeiro	40 (40 to 60)	25 (10 to 40)	6∞ (6 to 45)	3dia, 3 × 3 & 6∞
Hong Kong	30 (30)	18 (6 to 70)	8 × ∞ (6 × ∞ to 8 × ∞)	3 × 3 & 3 × ∞ (2 × 2 to 4 × 8; 1.5 × ∞ to 8 × ∞)
Tokyo	25 (25 to 30)	25 (20 to 25)	5∞ (5∞ to 20∞)	5∞ (2.5∞ – 30∞)
Sydney	50 (50)	40 & 15 (15 to 60)	30∞	4 × 6 & 5∞ (2 × 6 – 10 × 10) (5∞ – 7∞)
London	40 (40)	40 (20 to 60)	24∞	2 × 2, 2 × 3, & 2 × 6 (2 × 2 – 8 × 8)
Paris	38 (30 to 38)	18 (15 to 25)	6 dia (6dia) (3 × 3 to 6 × 6)	6 dia (6)
New York	40 (30 to 40)	25 (25 to 42)	7 ∞ (6∞ – 10∞)	4 × 6 (3 × 3 to 6 × 10)
Seattle	35 (25 to 38)	15 (12 to 36)	10∞ (8∞ – 10∞)	4 × 4 & 8∞ (4 × 4 to 5 × 5) (8∞ to 12∞)
Jacksonville	40 (30 to 40)	25 (10 to 38)	15∞ (15∞ to 24∞)	5 × 6 & 10∞ (3 × 3 to 6 × 6) (6∞ to 25∞)
Vancouver	40 (40)	30 (14 to 60)	2.5 × 6 (2.5 × 6)	8 × ∞ & 4 × 4 (3 × 3 to 2 × 8; 4 × ∞ to 12 × ∞)
Moscow	20 (6 to 25)	20 (6 to 25)	135∞ (35∞ to135∞)	15∞ (10∞ to 15∞)
Stockholm	28 (20 to 60)	30 (25 to 55)	20∞ (15∞ to 26∞)	3 × 4, 6dia, & 6∞ (3 × 3 to 4 × 4)
Beijing	10 (10 to 20)	15 (6 to 20)	3 × ∞ (3 × ∞)	3 × 3 & 3∞ (1× 1 to 3 × 3) (3∞ to 4∞)
Los Angeles	50 (20 to 120)	15 (15 to 38)	20∞ (4∞ to 32∞)	8∞ (8∞ to 12∞)
Athens	30 (15 to 38)	15 (10 to 22)	3 × 3 (3 × 3)	2 × 2 (1.5 × 1.5 to 3 × 3)
Tel Aviv	60 (15 to 65)	40 (10 to 70)	12 × ∞ & 6 × 6 (3 × 3 to 6 × 6; 6 × ∞ to 15 × ∞)	3 × 3 (2.5 × 2.5 to 8 × 10; 15 × ∞)
New Delhi	40 (10 to 60)	60 (20 to 60)	25∞ (25∞)	8∞ (4∞–30∞) (3 × 3 to 4 × 6)
Brasilia	40 (30 to 215)	30 (15 to 50)	20∞ (15∞ to 20∞)	3 × 3 & 30∞ (2 × 2–7 × 7) (10∞ to 35∞)
Lagos	50 (10 to 80)	20 (6 to 100)	12 × ∞ (8 × ∞ to 0 × ∞)	8 × ∞ (4 × ∞ to 30 × ∞)

(continued)

Table 6.1 (continued)

	Tree spacing (ft.)		Tree well (ft.)	
City	Blvd.	Street	Blvd.	Street
Johannesburg	70 (30 to 70)	40 (15 to 100)	12∞ (2∞ to 12∞)	3 × 3 & 10∞ (2 × 2 to 4 × 4) (2∞ to 18∞)
Tehran	20 (6 to 36)	12 (12 to 20)	6∞ (6∞ to 9∞)	6∞ (4∞ to 6∞)
Buenos Aires	35 (35)	30 (6 to 60)	50∞ (50∞)	4 × 4 (1 × 1 to 5 × 3)
Dubai	35 (15 to 50)	20 (18 to 25)	18∞ (10∞ to 30∞)	10∞ (10∞ to 15∞)
Phoenix	50 (18 to 100)	20 (20 to 50)	12∞ (8∞ to 25∞)	10∞ (6∞ to 30∞)
Cairo	40 (8 to 40)	15 (5 to 40)	10∞ (4∞ to 30∞)	2 × 2, 1dia,& 6∞ (1 × 1 to 4 × 10) (1dia to 4dia) (1∞ to 15∞)
Murmansk	6 (6 to 7)	20 (6 to 25)	25∞ (25∞)	15∞ (15∞ to 25∞)
Nuuk	No street or boulevard trees			
Reykjavik	na	20 (6 to 30)	na	3 × 3 (2 × 2 to 6 × 6)
Quito	25 & 50 (15 to 125)	22 (15 to 55)	10∞	1 × 1 & 1dia (1 × 1 to 4 × 4) (1∞ to 6∞)
Lhasa	10 (10 to 50)	15 (6 to 30)	6 × ∞ (4 × ∞ to 6 × ∞)	3 × 3; 6 × ∞; 1.5 dia. (2 × 2 to 6 × 6; 6 × ∞ to 25 × ∞′ 1.5 dia.)
Cusco	50 (8 to 160)	16 (4 × ∞ to 20 × ∞)	8 × ∞ (4 × ∞ to 20 × ∞)	8 × ∞; 2 dia. (2 × ∞ to 10 × ∞; 2 dia.)

∞ = continuous planting strip; [1] modal tree spacing; [2] range in tree spacing

safety feature. Wider spacing avoids the "palisade effect" that occurs when a row of closely planted trees are viewed from a narrow angle. The "palisade effect" can prevent motorists from seeing cross traffic or pedestrians approaching at intersections. The modal distance for the spacing of trees in boulevard median was 40 feet for the 33 cities in the study, the average distance was 34 feet. In contrast the modal and average distances measured for street trees were 20 and 23 feet. The closer spacing generally used for street trees is a reflection of the potential size of the tree crown at maturity. A continuous crown canopy above the sidewalk is desired for shading pedestrians in most cities. The reduced growing space along streets, in contrast to the space available in most boulevard medians, dictates the use of smaller trees.

Design precedents play a part in the selection of tree spacing along streets and boulevards. It is interesting to note that the same spacing of boulevard trees that was used in London shows up in several cities that were once English colonies, while this "British spacing" is not found in other European cites nor in cities outside of Europe that were not previous British colonial cites (Table 6.2).

Table 6.2 Comparison of modal spacing of boulevard trees in selected cities

City	Modal Boulevard tree spacing (ft.)
London	40
New York[a]	40
New Delhi[a]	40
Cairo[a]	40
Singapore[a]	40
Vancouver[a]	40
Sydney[a]	50
Johannesburg[a]	70/35[b]
Paris	38
Seattle	35
Athens	30
Hong Kong[a]	30
Stockholm	28
Tokyo	25
Moscow	20
Tehran	20
Beijing	10
Lhasa	10

[a]Former British Colonial City;
[b]70 feet commonly used for palms while 35 feet is used broadleaf species of shorter stature

The difference observed in the spacing of boulevard trees in tree dominated biomes compared to biomes not dominated by trees was not parallel at the level of street trees. The average modal street tree spacings were 27 and 25 respectively for the two types of biomes.

Planting strips, tree wells, and sidewalk cuts for tree planting varied considerably among the cities in the study. The modal boulevard-median planting strip width was six feet. The range of modal widths varied from 5 to 135 ft. In two cities in the study the median strips were always paved and tree wells either three by three feet or six feet in diameter were used. The average of the modal median widths for tree dominated biomes was 14 feet (when the unusually wide medians used in Moscow [135 feet] were excluded from the average), while those of the non-tree dominated biomes were 19 feet. This difference may be related to the need to provide more growing space to larger trees used in boulevard medians in non-tree dominated biomes.

Planting spaces for street trees can be categorized as strips, squares and rectangular sidewalk cuts, circular tree wells, and containers. The modal width of planting strips in the 33 cites studied was 6 feet. Modal planting strips ranged in width from 3 to 30 ft. The modal rectangular or square sidewalk cut for tree planting was three by three feet. The size of these cuts varied from one by one foot to eight by eight feet. The modal circular tree well was six feet in diameter, with a range from one to six feet in diameter tree wells. No particular pattern was evident with regard to the

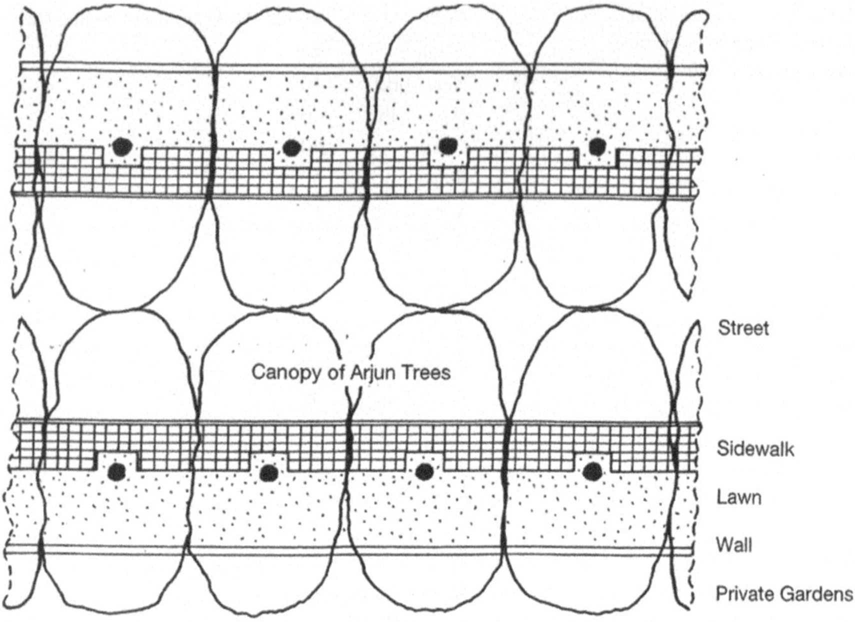

Fig. 6.2 Planting pattern along Nini Marg Street, New Delhi

spaces provided for street tree planting among the biomes. It was somewhat discouraging to see that small planting spaces were being used in many cities. Many of these smaller planting strips, sidewalk cuts, and tree wells showed evidence of sidewalk damage by tree roots and injury to the trees by the restrictions of the growing space. London plane trees growing in sidewalk cuts less than four by four feet caused damage to the adjacent sidewalks and curbs in Palo Alto, California (McBride 2003). Studies by Urban (1989) suggest that six by six feet sidewalk cuts are necessary to sustain long-term tree growth.

Some unique examples of the patterns of street tree design and planting are presented in the following paragraphs. These represent variations in the typical pattern of single, straight rows of trees planted equal distant along streets observed in most cities.

New Delhi

Along Niti Marg in New Delhi arjun trees (*Terminalia arjuna*) are planted on 32 foot centers in notched planting spaces at the edges of sidewalks (Fig. 6.2). This pattern informs the pedestrian of the presence of the tree because of its partial blocking of the view along the inner edge of the sidewalk. It also requires a slight "walking around" the tree if one is moving along the inner edge of the sidewalk. The

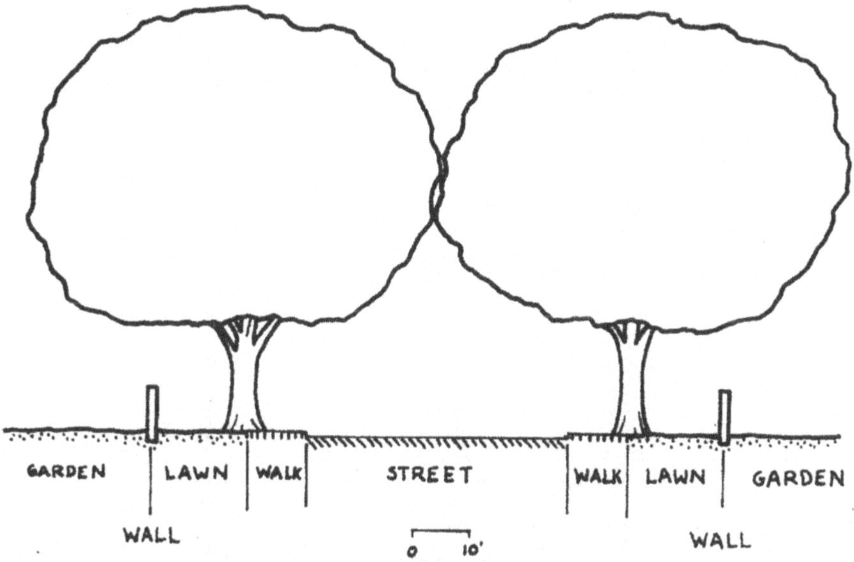

Fig. 6.3 Cross-section of Net Marg street, New Delhi

design brings the tree into the experience of walking along the street in a way that is not obtrusive. The evergreen arjun trees planted along Niti Marg have been allowed to extend their canopies out to the middle of the street and over the fifteen feet wide lawn adjacent to the sidewalk (Fig. 6.3). The shade created by the trees is appropriate to the savanna/scrub biome in which New Delhi is located. The shade and the buttressed roots of the arjun trees adds to the inviting nature of the street. The design was created by Lutyens and Baker who employed it along many residential streets in New Delhi (Irving 1981).

Stockholm

Shade is not a premium design characteristic of streets in the mixed deciduous-coniferous forest biome city of Stockholm. The northern latitude location of Stockholm makes the warmth of sunlight valuable during all of the months of the year. Only on summer afternoons will shade be appreciated along streets in the city. A design that enables sunlight to reach the sidewalk during the morning, but shades it during the afternoon was observed along Timmermansgatan Street in Stockholm. Linden (*Tilia cordata*) had been planted on 50 feet centers in a single row along the west side of this north-south oriented street (Fig. 6.4). The trees have been pruned up to a height of ten feet and maintained with a crown radius of fifteen feet. Rays of the morning sun in the summer come in under the canopy to warm the sidewalk and pedestrians along the west side of the street. By noon tree canopies shade the

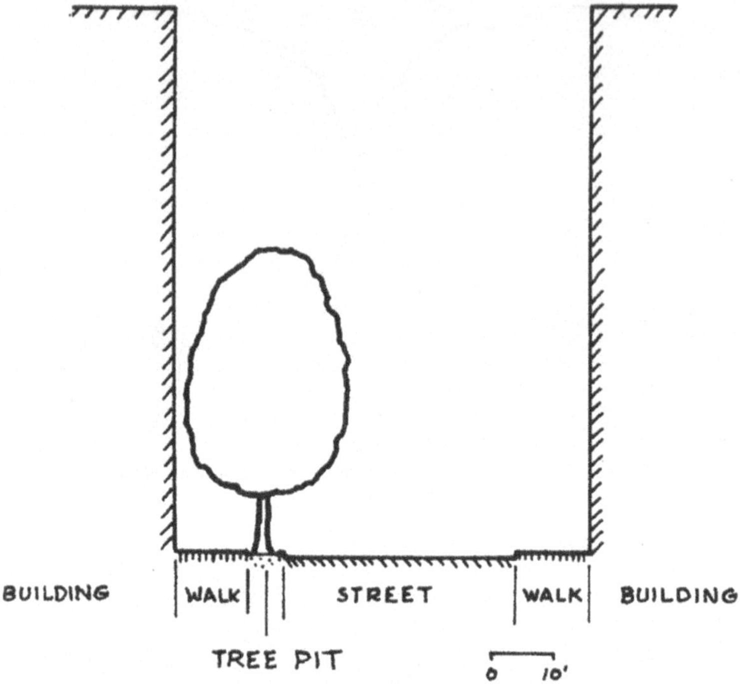

Fig. 6.4 Cross-section of Timmermansgatan Street, Stockholm

sidewalk to lower the temperature and by mid-afternoon their shadows reach the sidewalk on the other side of the street. It is a very simple design solution that recognizes the need for warmth in the morning in a northern city, even in the summer.

Murmansk

Moving further north one finds an even greater need to approach street tree design from a functional standpoint (McBride and Doughovnikoff 2011). Murmansk, located above the Arctic circle in the tundra biome, experiences low temperatures throughout the year. Minimum temperature have fallen below –75 °F and below freezing temperatures can occurred in every month of the short summer. Strong winds create "wind chill" temperatures that are uncomfortable during the summer and dangerous during the winter. To provide protection for pedestrians along the major streets, like Lenin Avenue, three to five rowan (*Sorbus acuparia*) seedlings were planted on 5–7 foot centers in a planting strip between the sidewalk and the street (Fig. 6.5). As they grew they were pruned to prevent their lower branches from interfering with the movement of pedestrians and vehicles, but no pruning has been done between the trees along the planting strip. The result is a wall of branches

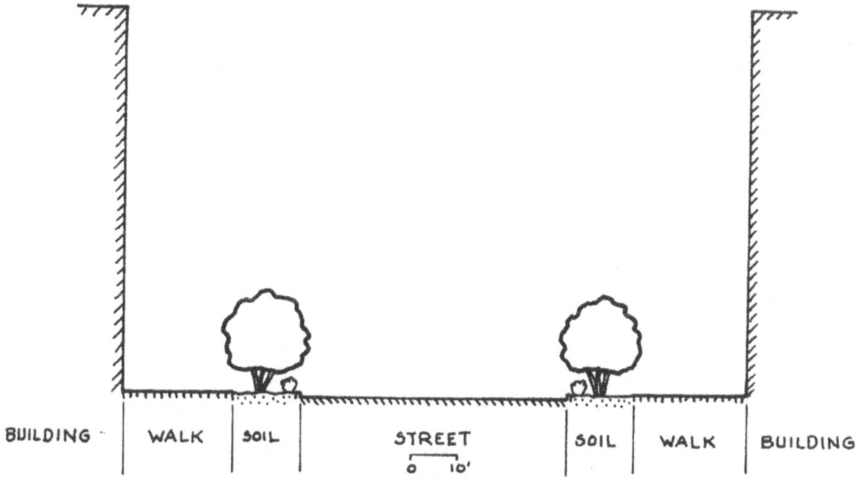

Fig. 6.5 Cross-section of Lenin Avenue, Murmansk

and foliage to a height of about 25 feet. This hedge-like row of street trees reduces the wind velocity in both the winter and the summer. It also creates a sense of protection from the vehicles moving along Lenin Avenue. The slight attenuation in traffic noise adds to the pedestrian environment created by the tight spacing of the trees.

Seattle

Typical patterns of trees along streets and boulevards are linear rows in tree pits or planting strips along the margins of sidewalks (Arnold 1993; Jacobs et al. 2002). This arrangement provides minimum interference with pedestrians moving along sidewalks, but may interfere with drivers seeing on-coming traffic at intersections and pedestrians seeing people exiting from parked automobiles (Jacobs 1993). An alternative arrangement for street tree design can be seen along Second Avenue in Seattle. Here western red cedar (*Thuja plicata*) has been planted in a row along the middle of the 20 foot wide sidewalks (Fig. 6.6). In 1999 these trees averaged eleven inches in diameter and thirty feet in height. They were planted on 32 feet centers. The dramatic effect of a row of conifers down the middle of the wide sidewalk was impressive. The choice of western red cedar is in keeping with coniferous forest biome in which Seattle is located.

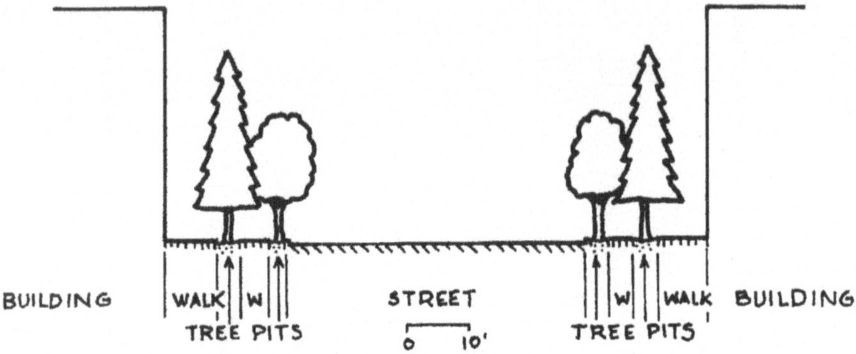

Fig. 6.6 Cross-section of Second Avenue, Seattle

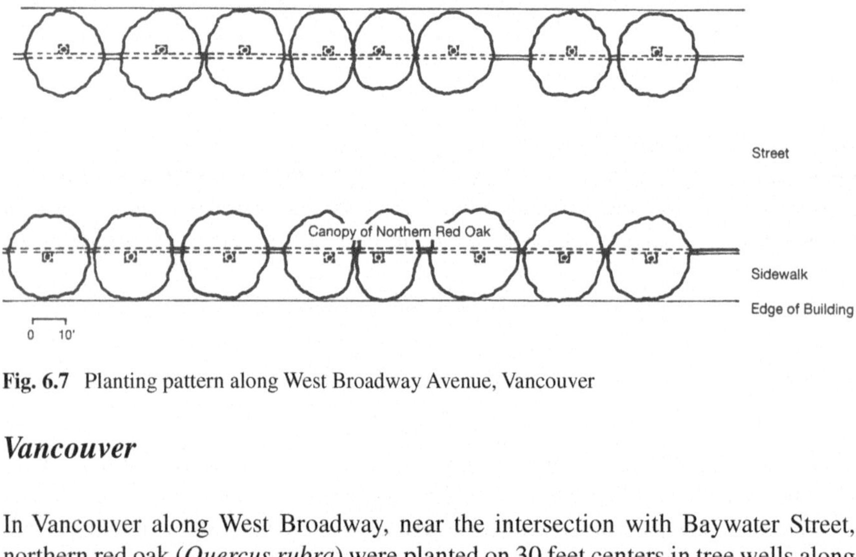

Fig. 6.7 Planting pattern along West Broadway Avenue, Vancouver

Vancouver

In Vancouver along West Broadway, near the intersection with Baywater Street, northern red oak (*Quercus rubra*) were planted on 30 feet centers in tree wells along the edge of the sidewalk. This pattern appears quiet common until one approaches the middle of the block where the spacing between two adjacent trees is only 15 feet (Fig. 6.7). This change in spacing breaks the rhythm of the design. The pattern of halving the tree spacing near the middle of the block is repeated for several blocks along West Broadway and on other streets. The effect reminds one of how a rhythm is established by tree spacing along a street. The breaking of this rhythm is somewhat disturbing, almost unnoticed, but in subsequent blocks one is challenged to find the break in tree rhythm as one drives or walks along the street.

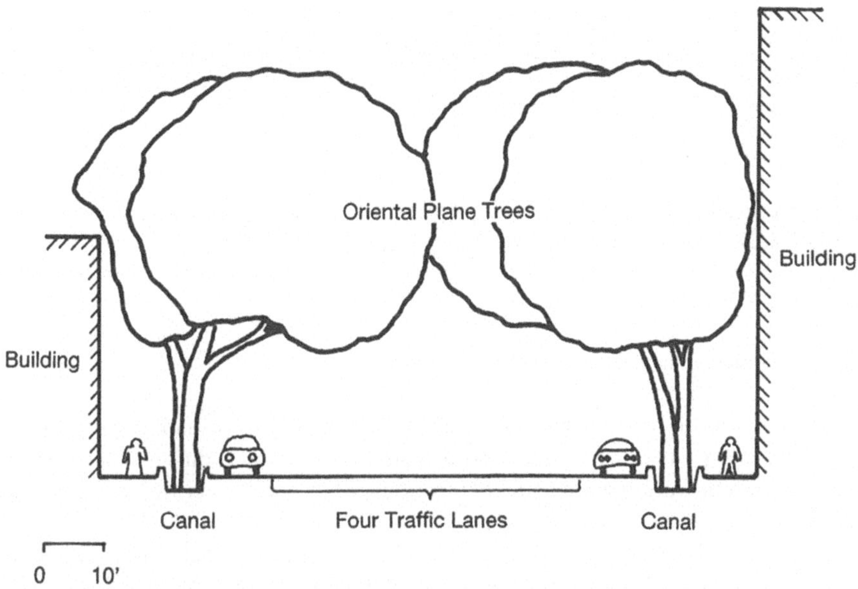

Fig. 6.8 Cross-section Valisar Avenue

Tehran

Tree have been planted in canals along the major avenues in Tehran. A series of avenues, which have canals along both sides of the road, cross the city from its higher elevations at the base of the Alborz Mountain to its lower elevations. These avenues and their canals were constructed during the early part of the Pahlavi Dynasty (1921 to 1979) when Reza sought to modernize the country and convert Tehran into a modern city (McBride and Mossadegh 2000). These efforts involved the creation of the grid system of avenues and boulevards to facilitate the movement of automobiles within the city.

The unique feature of the new avenues established during the early Pahlavi Dynasty were the canals constructed along their edges. Twelve avenues were chosen for canal construction. The canals served both as a place to plant trees and a means of irrigating them. Typically six feet (2 m) wide and three feet (1 m) deep, these canals were planted with either one or two rows of Oriental plane trees on three to five feet (one to 1.5 m) centers (Fig. 6.8). The floors of the canals, except for small tree wells where trees have been planted, were paved with stream cobbles. Four to eight inch (10 to 20 cm) high walls were constructed at intervals across the canals to decrease the velocity of the flowing water. The distance between the walls varied with the gradient of the canals, in some places being only a few meters apart.

The canals are flooded to irrigate the trees by introducing water from reservoirs or ghanats at the upper elevation edge of the city. Water is released into the canals in the morning and is allowed to run in the canals, at a depth of two to four inches

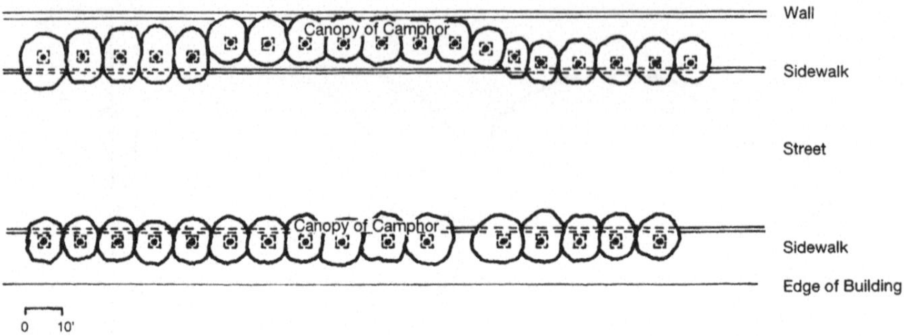

Fig. 6.9 Planting pattern along Evergreen Road, Ningbo

(5–10 cm), until it reaches the lower end of each canal. This usually requires four hours for most canals. The trees are irrigated once a week during the summer months, except during especially hot periods when two waterings a week are required.

In the decades since the construction of the tree-lined canals, the trees have grown to heights of 20–30 m and diameters of 30–50 cm. They now shade the avenues with dappled light, reduce the temperature of the urban heat island, and eliminated the glare of the summer sun. Water moving down the canals adds to the experience of pedestrians along these avenues. The motion of the water, its fresh odor, and the sound it makes falling over the check dams add to the character to these special avenues.

Other Cites

Two other unique variations in street tree design that were observed during travels associated with urban forests and air pollution research in China. In Ningbo camphor trees (*Cinnamomum camphora*) have been planted on ten feet centers along the edge of Evergreen Road, except where the sidewalk is widened to the edge of a canal in a gap between the shops lining the street. Here the row of camphor trees is gradually curved in to the middle of the widened sidewalk (Fig. 6.9). This graceful curve of trees calls the pedestrian's attention to the canal. One can sees people moving with the curving row of trees to inspect the canal, while others filter through the trees to continue on their straight path. Like the simple change in spacing along West Broadway in Vancouver, this simple change in the pattern of trees is a delightful alternative to the usual rigidity of the linear row. It has interesting applications in other cities where it is desired to direct the attention of pedestrians to features of special interest away from the street.

The city of Kunming has is a unique street with an adjacent sidewalk that was built through a eucalyptus plantation. The designers incorporated two rows of eucalyptus trees from the plantation into the sidewalk by creating a series of "islands" of

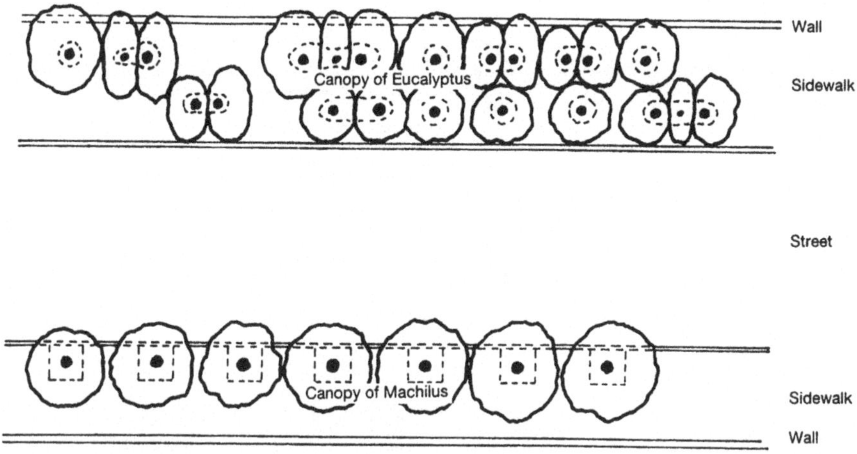

Fig. 6.10 Planting pattern along University Drive, Kunming

one, two, or three trees (Fig. 6.10). The effect is that of an irregular allee of eucalyptus trees that seem to be constantly changing as a pedestrian walks down the sidewalk.

Traffic Barriers

Traffic barriers are located in some cities at street intersections to reduce traffic speed without causing traffic to come to a stop. Barriers are also installed to prevent through traffic from using residential streets. Planting designs for traffic barriers varied with the size, shape, and function of the barrier. Traffic circles and barriers to close access to streets were observed in five of the cities in this study: Dubai, New Delhi, Quito, Seattle, and Sydney. Traffic circles were common in the first four of these. Barriers used to close streets to through traffic were common in some neighborhoods in Sydney.

Traffic circles in Dubai, New Delhi, and Quito were employed at the intersections of major avenues, while traffic circles were used only in residential areas on secondary streets in Seattle. This difference in location, relative to street size, resulted in a large difference in the size of traffic barriers in Dubai, New Delhi, and Quito in contrast to Seattle.

The modal diameter of traffic circles in Dubai, New Delhi, and Quito was 200 feet, which provided 31,400 feet^2 of space for planting. In Seattle the modal diameters of the traffic circles was 15 feet providing only 177 feet^2 of planting space. In Dubai, New Delhi, and Quito the number of trees counted in traffic circles ranged from 1 to 34. An average of 9 trees were observed in traffic circles. This average is skewed by a traffic circle in New Delhi that supported 34 trees. The modal number of trees found in traffic circles in the three cities was 7. The aver-

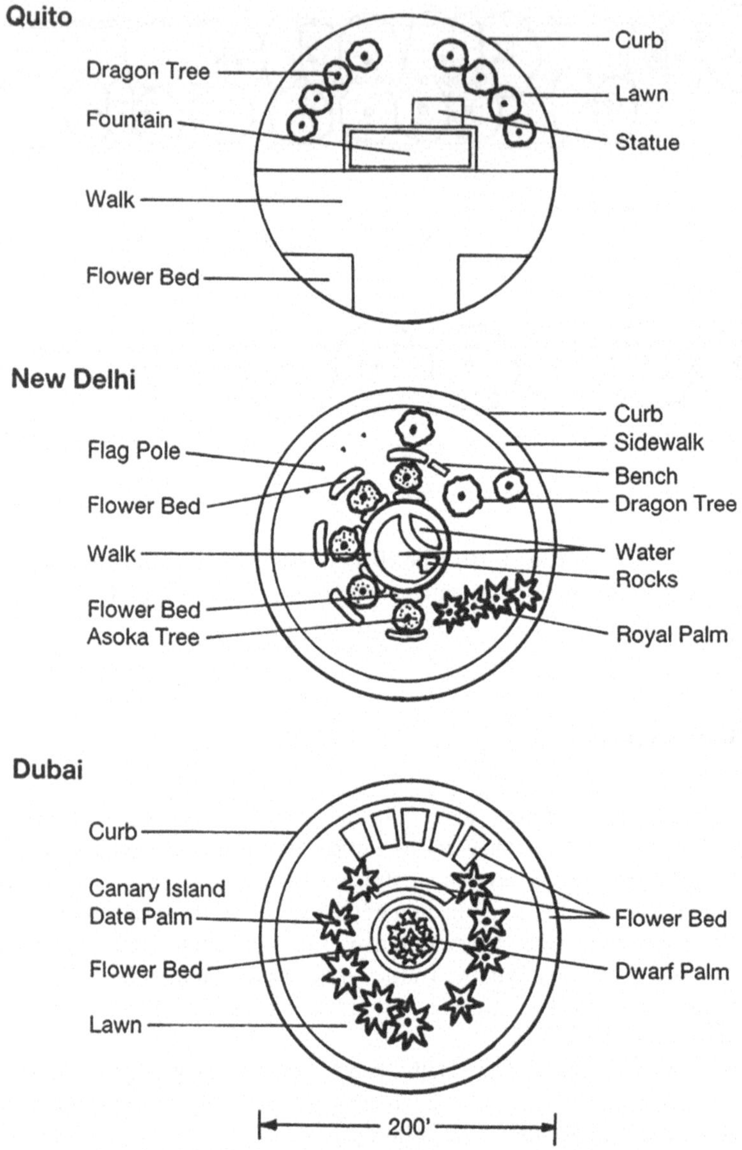

Fig. 6.11 Planting patterns in three large traffic circles in Quito, New Delhi, and Dubai

age tree density for the three cities was 4 trees/0.25 acre (10,890 feet^2). Average tree canopy cover was 14%.

Considerable variation in the pattern of tree planting was observed on large traffic circles in Dubai, New Delhi, and Quito (Fig. 6.11). Seventeen patterns were observed, only two of which occurred in more than one city. These two patterns are referred to as "interior circle of trees" and "interior arc of trees". Both of these patterns were observed in both Dubai and Quito. All of the traffic circles sampled in New Delhi, which had been planted with trees, exhibited different tree planting patterns. In New Delhi, garden clubs and civic groups have been given permission to design and maintain the plantings on large traffic circles. This may explain the variety of patterns observed in the design of New Delhi traffic circles. In Dubai and Quito large traffic circles were designed by the municipal horticulture departments and were maintained by the city. Traffic circles in all three cities were planted with shrubs, ground covers, grass, and often flower beds. Many large traffic circles were surrounded by sidewalks and in some cases dissected by sidewalks. Fountains, statues, art objects, and sculptures adorned many of the large traffic circles. Tree were often used on these traffic circles as a backdrop for the architectural features. Trees on large traffic circles were pruned to not interfere with a driver's ability to see vehicles entering or leaving the traffic circle at the adjacent intersection. Designs did not, however, always ensure an unobstructed line of sight across the large traffic circle.

The modal number of trees in the small traffic circles observed in Seattle was one and the modal tree density was 2 trees/100 feet^2. Tree canopy cover over the traffic circles in Seattle averaged 20%. Twelve patterns of tree planting were found in the small traffic circles in Seattle (Fig. 6.12). Seven of these were employed in the design of small traffic circles and five in oblong traffic barriers. Almost equal percentages of the "single, tree centered" (35%) and the "single tree off centered" patterns are used in the small circles. No single pattern was commonly used in the oblong barriers. The conditions of small traffic circles in Seattle varied in terms of the design and maintenance of the plants. Seventeen percent of the small traffic circles had been covered with asphalt and another 12% supported only weedy plants. These occurred in low income neighborhoods. In contrast many of the small traffic circles exhibited carefully-cared for flower beds, ground cover, shrubs, and sculptural pieces. Neighbors living near these small traffic circles appeared to have taken responsibility for their maintenance and design improvements. They were sites of pride for local neighborhoods and provided additional space for gardeners to work with plants.

The traffic barriers used in Sydney to close off one end of a residential street were rectangular in shape. They were created by blocking off of a portion of a street adjacent to an intersection. Their presence prevents vehicular access the closed-off street and to provide a relaxing setting for local residents. These traffic diverters ranged in size from 15 by 36 to 24 by 30 feet. The modal size had an area of 720 feet^2. Usually only two or three trees were planted on each of the tree barriers. The modal number observed was three trees. Modal tree density was 1 trees/200 feet^2. An average of 90% tree canopy cover was recorded over these traffic barriers. Trees were planted in circular tree pits surrounded by low walls topped with a surface designed for sitting (Fig. 6.13). These tree pits were positioned to prevent the passage of vehicles across the barrier. On narrower streets only two trees were required, while on wider streets three were used in an arc or triangular arrangement. Many of the traffic diverters in Sydney function as mini-parks.

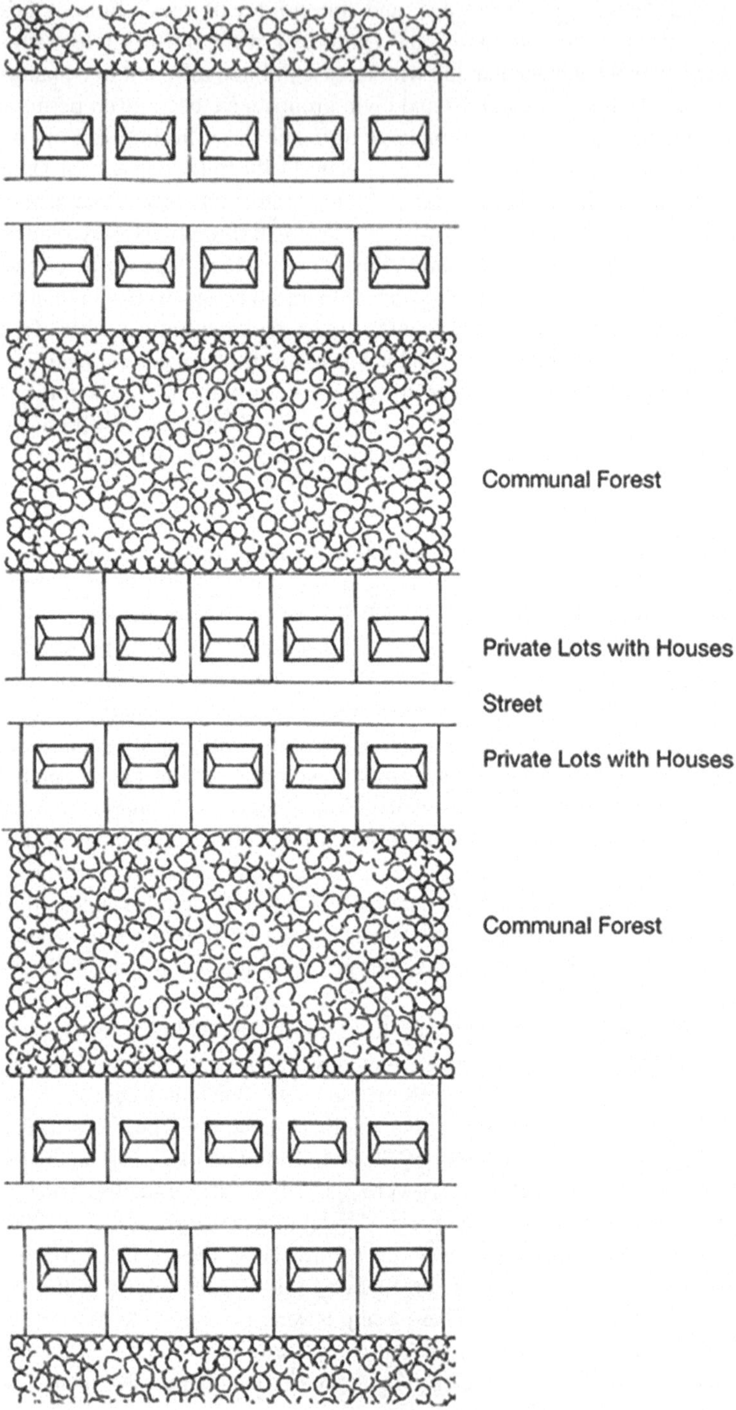

Fig. 6.12 Pattern of forest strips between rows of housing, Stockholm

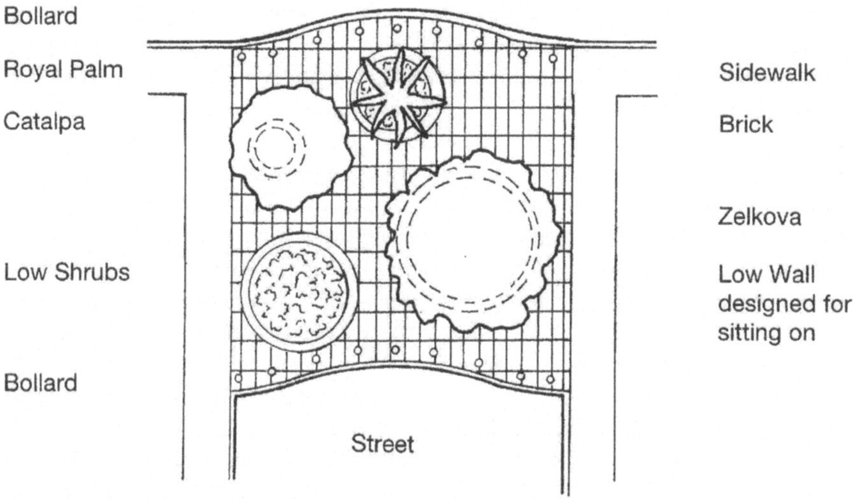

Fig. 6.13 Traffic diverter, Sydney

Although no cities in this study were located in the Netherlands, it is appropriate to mention the used of trees in several Dutch cities to calm traffic in specially designed residential streets called "woonerfs". These streets are designed to be shared by children and low volumes of vehicular traffic. Traffic speed is reduced in woonerfs by the strategic location of a clearly marked entry and trees and planting beds. Tree placement requires a bending of the driving lane that further reduces the speed of a vehicle. Sitting areas and benches are employed in woonerfs to enhance the use of these shared streets. Woonerfs have been constructed in several countries in addition to the Netherlands and may be observed in England, Sweden, Denmark, France, Switzerland, and Japan (Southworth and Ben-Joseph 1997). They are also employed in some residential areas in Sydney.

Small Parks

The small parks observed in this study usually occupied a city block and were generally less that 2.5 acres (1 ha) in size. The designers of these parks were not memorialized by bronze plaques in the parks, but were city staff landscape architects, engineers, or horticulturalists whose designs were constrained by economics, maintenance, and utilitarian considerations. Within these constraints they designed small parks that were both functional and attractive. Moreover, the small parks they designed were used by neighborhood people who seemed very happy with their parks. Small parks make a significant contribution to urban forests, which is large in proportion to the size of these parks. Small parks are often the first place that children in many cities are introduced to trees. Trees in these parks are an important part of the early childhood experience of nature and the initial bonding of many urban dwellers with trees (Cooper Marcus 1995).

One hundred and fifty-seven small parks were surveyed in the 33 cities visited in this study. Average tree canopy cover in these small parks ranged from <1 to 75% (Table 6.3). The lowest averages occurred in the desert and tundra biomes, while the highest average was recorded in the broadleaf evergreen forest. Small parks in forested biomes averaged 47% canopy cover and in non-forested biomes 28%. The lower percent canopy cover in the non-forested biomes may in part be due to the necessity of irrigation in many of these biomes. Tree irrigation is a costly investment and may have kept down the number of trees planted in small parks in these biomes. Canopy cover in small parks is also due to the use patterns of small parks. The low percent canopy cover in the coniferous forest biomes cities of Jacksonville and Seattle was due to the use of parks for softball and soccer fields. Forty-four percent of the small parks in these cities had perimeters of trees with the interior portions set aside for ball playing. One third of the small parks in cities in the savanna-thorn scrub biome and one fourth of the parks in cities in the desert biome were devoted to soccer fields.

In the desert biome, one might expect to see a higher percent canopy cover in small parks to provide shade; however, the lowest average percent canopy covers were measured in desert cities. Low percent canopy cover was maintained in these desert cities, according to local urban foresters, to allow for rapid re-radiation of heat from the ground surface in the early evening. Parks are primarily used in the evening in these desert cities because of the excessive heat during the day. Many parks in Cairo and Dubai have street lights over areas of lawn to facilitate park use at night.

Sixteen patterns of tree planting were observed in the parks (Fig. 6.14, Table 6.4). These ranged from a single tree to complex patterns involving perimeter plantings combined with small groves of trees. The most commonly encountered pattern was a perimeter planting of trees with the interior of the park planted in lawn and sometimes flower beds. This pattern provides marginal shade while keeping the interior of the park open for activities suited for lawns. The perimeter pattern also serves to delineate the boundary of the park and to provide some screening from adjacent sidewalks, street and buildings. The percentage of parks where the perimeter pattern was used ranged from 0 to 50% of the cities of various biomes. It was not observed as a pattern in small parks in the desert or tundra biomes, but was used in one or more cities in all of the other biomes. The absence of perimeter planting in small parks in the desert biome is related to the needs to reduce tree canopy cover in order to promote re-radiation of heat in the evenings when park use its at a maximum. In the Tundra biome the grove pattern was used to provide wind protection within small parks. The grove pattern used in the tundra biome was the second most frequently observed pattern in other biomes as well. The grove pattern is characterized by one or more small groups of trees planted in a park without any perimeter or other arrangement of trees. Seven of the eleven biomes had cities in which the grove pattern had been used. A similar frequency of occurrence was observed for the linear pattern. This pattern is a modification of the perimeter pattern in which a linear row

Table 6.3 Tree canopy cover and facilities in small parks

Biome/City	Number of samples	Canopy cover (%)	Facilities[a]			
			Totlot	Playfield	Basketball	Tennis
Tropical forest						
Bangkok	4	49	50	0	25	0
Rio de Janeiro	8	69	52	0	0	0
Singapore	4	46	75	0	0	0
Average	5	55	59	0	8	0
Broadleaf evergreen forest						
Sydney	2	50	0	0	0	0
Tokyo	5	74	40	20	0	0
Hong Kong	6	32	50	17	0	0
Average	4	52	30	12	0	0
Deciduous forest						
London	10	44	30	0	0	0
New York	3	48	67	0	0	36
Paris	4	49	25	0	0	0
Average	6	47	41	0	0	12
Coniferous forest						
Jacksonville	10	11	33	50	0	0
Seattle	3	25	0	33	0	0
Vancouver	6	12	67	50	0	50
Average	6	16	33	44	0	17
Mixed deciduous-coniferous forest						
Beijing	2	75	50	50	50	0
Moscow	8	43	12	0	0	0
Stockholm	4	38	25	0	0	25
Average	5	52	29	17	17	8
Mediterranean woodland scrub						
Athens	3	43	0	0	0	0
Los Angeles	5	29	60	0	50	20
Tel Aviv	9	38	76	0	0	0
Average	6	37	45	0	17	7
Savanna/ Scrub						
Brasilia	3	70	0	33	0	0
New Delhi	4	5	0	100	0	0
Lagos	3	40	0	0	0	0
Average	3	38	0	33	0	0
Grassland						
Buenos Ares	6	32	17	0	0	0
Johannesburg	7	35	53	0	0	14
Tehran	3	42	67	0	0	0
Average	5	36	46	0	0	5

(continued)

Table 6.3 (continued)

Biome/City	Number of samples	Canopy cover (%)	Facilities[a] Totlot	Playfield	Basketball	Tennis
Desert						
Cairo	4	12	50	0	0	0
Dubai	7	8	28	0	0	0
Phoenix	4	9	0	80	0	0
Average	5	10	26	27	0	0
Tundra						
Murmansk	3	23	0	0	0	0
Nuuk	1	<1	0	0	0	0
Reykjavik	2	4	0	0	0	0
Average	2	9	0	0	0	0
Highlands						
Lhasa	3	30	0	0	0	0
Quito	8	16	0	0	0	0
Cusco	7	9	11	0	11	0
Average	6	18	4	0	4	0

[a]Percent of samples with each facility

of trees is planted along one side of the park. It was often observed in small parks that were bounded on three sides by buildings. In these cases the linear row of trees were planted along the street side of the park.

Large Parks

Large parks are characteristic of the world's great cities. These parks serve as important landmarks in many cities. New York City's Central Park and the Bois de Boulogne in Paris are as much a part of local people's images of these cities as are the Empire State Building and the Eiffel Tower. Large urban parks present a great richness of tree species and provide a variety of settings in which trees can be exhibited and viewed. Trees in large urban parks provide wildlife habitat, reduce air pollutants, and ameliorate urban heat islands. Their size also allows space to serve a variety of recreational objectives of urban dwellers and provide for large outdoor gatherings.

A sample of three or four of the largest urban parks was examined in each of the cities surveyed. The intent was not to complete a quantitative inventory of trees in these parks, but to take notes on the general design characteristics of each park and the kinds of trees observed. The large parks, excluding the large parks set aside primarily as nature reserves (e.g., the 29,650 acre Parque National da Tijuca in Rio de Janeiro and the 3706 acre Griffith Park in Los Angeles), ranged in size from the 27 acre Kungliga Humlegarden in Stockholm to the 2,457 acre Bois de Vincennes

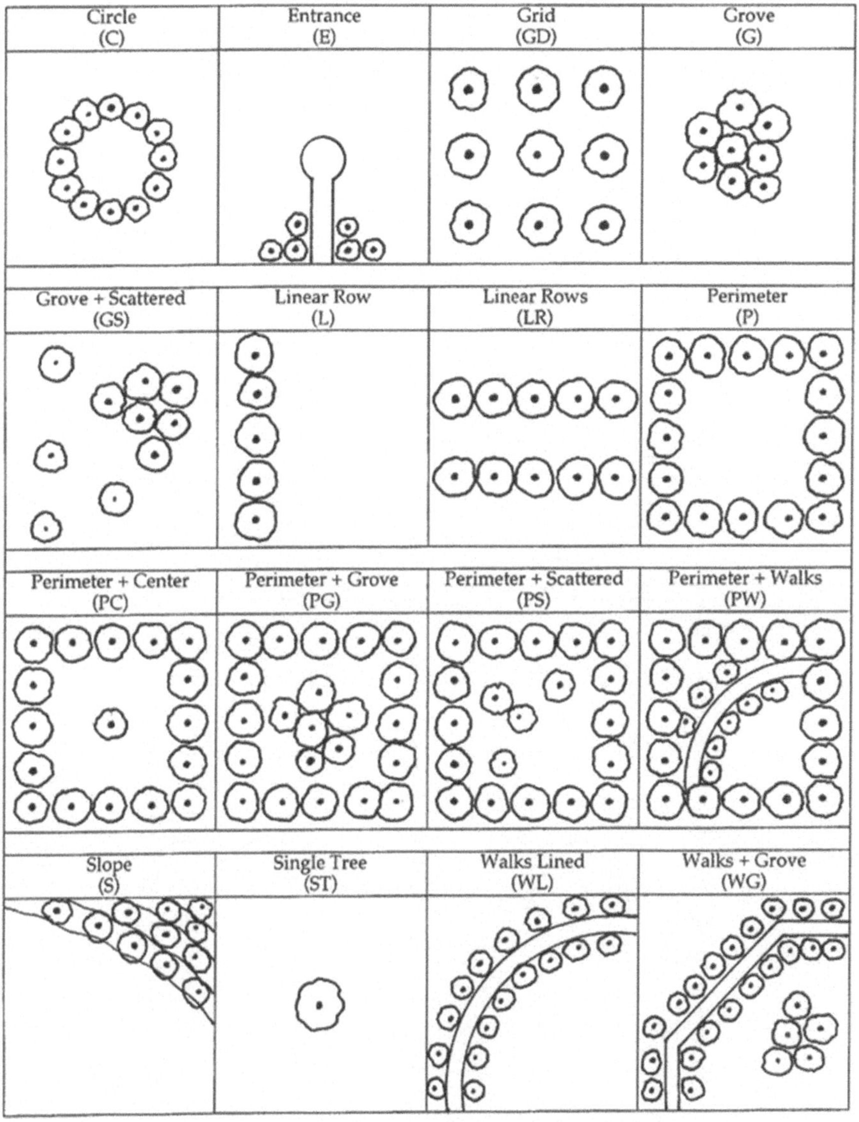

Fig. 6.14 Patterns of tree plantings observed in small parks

in Paris. The average size of the largest park in each of the cities studies was over 400 acres (162 ha), when the very large nature preserves were not included in the average. The canopy cover ranged from a low of less than 1% in twentieth century activity oriented parks to nearly 100% canopy cover in nature parks. Tree species richness increased from five to over 300 species as one moved from twentieth century activity parks to nature parks in the forested biomes.

Table 6.4 Occurrence and overall percentage of tree planting patterns observed in small in different biomes

Biome/City	Tree planting pattern[a]															
	C	E	GD	G	GS	L	LR	P	PC	PG	PS	PW	S	ST	WL	WG
Tropical forest			•	•				•		•						•
Broadleaf evergreen forest						•		•	•	•						
Deciduous forest			•	•			•	•		•	•	•	•			
Coniferous forest				•	•	•		•	•	•	•					
Mixed deciduous coniferous forest	•			•		•		•	•	•		•				
Mediterranean woodland scrub						•		•								
Savanna/Scrub					•			•								
Grassland				•	•	•		•			•	•			•	
Desert		•		•		•			•	•		•		•	•	
Tundra				•		•		•				•				
Highlands		•		•		•	•	•	•	•		•				
Overall average	0.7	1.2	1.4	8.4	6.5	11.1	1.3	34.5	4.6	15.1	1.7	15.6	0.4	0.6	1.5	1.5

[a]*C* circle, *E* entrance, *GD* Grid, *G* Grove, *GS* Grove + Scattered, *L* Linera row, *P* Perimeter

Large urban parks may be divided into six categories on their basis of their origin and use: (1) former royal gardens, (2) former hunting parks, (3) nineteenth century parks, (4) terrain defined parks, (5) nature parks, and (6) twentieth century parks. Examples of parks in these categories for the cities surveyed in this study are shown in Table 6.5. These categories are defined in the following paragraphs. It should be pointed out that the categories are arbitrary in that some former royal gardens were redesigned both in the nineteenth and twentieth centuries to better fit the park concept of the time. Also some areas that could not be economically developed for housing and other urban uses in the nineteenth century, because of terrain, were later designated as nature parks.

Large gardens were built adjacent to royal palaces throughout Europe during the Renaissance. Many of these were within or at the edges of existing cities at the time of their construction. These gardens were characterized by formal designs in which trees were used to line walks or to create small bosques (Lawrence 1995). Large urban parks derived from royal gardens are generally characterized by a lower percentage of tree canopy cover than parks derived from royal hunting parks or parks built during the nineteenth century. One finds interesting exotic tree species in these former royal gardens that were planted as specimen trees for the former monarchs. These royal estates and their gardens were surrounded by urban growth as European capital cities grew during the eighteenth century. During the late eighteenth and early nineteenth century public pressure caused monarchs to open their gardens to the public. Kensington Garden Park in London, for example, was the garden adjacent to Kensington Palace while it was occupied by the British monarchs, William and Mary beginning in 1688. It was subsequently combined with the adjacent Hyde Park to form one of the largest public open space areas in London. In Kensington Garden Park portions of the original formal royal garden adjacent to the Orangery remain intact, while other portions of the former royal garden have been modified to accommodate public use. Tree canopy cover in Kensington Garden is about 35 percent and is comprised of over two dozen tree species.

Large hunting estates were also established in the vicinity of European Renaissance cities as game parks for hunting by royalty and wealthy aristocrats. These were generally created by tree planting on agricultural lands, although some were carved out of existing forests. A large percentage of tree canopy cover was desired to provide habitat primarily for deer in these parks. The typical layout of a hunting park included the fanning out of avenues through the planted forest. Hunters would walk or ride along these avenues while a beater drove game from the forest on to the avenues, where they would be shot. Sokolniki Park is an example of a former hunting park that now serves as an important urban park in Moscow. It still maintains the avenues laid out for hunting purposes when it was part of the Sokolniki estate outside of the city wall in the seventeenth century. Like other large urban parks derived from hunting parks in Europe it supports a very high percentage of tree canopy cover. The tree canopy cover of Sokolniki Park is approximately 85%. Former hunting parks have a large percentage of native tree species, but are not generally characterized by high species richness. Soloniki Park supports a number of tree species common to the mixed deciduous-conifer forest biome in the vicinity of Moscow, but relatively few exotic species.

The middle of the nineteenth century was a time when new large parks were built in cities in many parts of the world. These new urban parks were inspired by the

Table 6.5 Examples of different categories of large parks

Park category	Biome	City	Park
Nature Park	Tropical forest	Rio de Janerio	Parque Nacional da Tijuca
		Singapore	Bukit Timali Nature Preserve
	Broadleaf evergreen forest	Sydney	Botany Bay National Park
	Coniferous forest	Jacksonville	Tree Hill Park
		Seattle	Discovery Park
		Vancouver	Stanley Park
Terrain Defined Park	Tropical forest	Rio de Janerio	Sugar Loaf
	Deciduous forest	New York	Van Courtland Park
	Mixed deciduous-conifer forest	Moscow	Discovery Park
	Mediterranean woodland-scrub	Athens	Likavitos Hill
		Los Angeles	Griffith Park
Former Royal Gardens	Broadleaf evergreen forest	Tokyo	Royal Garden Park
	Deciduous forest	London	Kensington Garden/ Hyde Park
			Regents Park
		Paris	Jardin du Luxemburg
	Mixed deciduous-coniferous forest	Stockholm	Hagaparken
	Highlands	Lhasa	The Norbulingka
Former Hunting Parks	Coniferous forest	Moscow	Sokolniki Park
	Mixed deciduous-coniferous forest	Stockholm	Djurgarden
Nineteenth Century Parks	Broadleaf evergreen forest	Sydney	Hyde Park
	Coniferous forest	Seattle	Volunteer Park
	Mixed deciduous-coniferous forest	Stockholm	Kungliga Hamligarden
	Mediterranean woodland-scrub	Los Angeles	Elysian Park
	Undifferentiated highlands	Quito	Parque El Ejido
Twentieth Century Parks	Tropical forest	Brasilia	Parque da Cidade
	Broadleaf evergreen forest	Hong Kong	Hong Kong Park
	Conifer forest	Jacksonville	Orange Park
	Mediterranean woodland-scrub	Tel Aviv	HaYarkon Park
	Savanna-thorn scrub	New Delhi	Buddha Jayanti Park
	Grassland	Tehran	Mellat Park
	Desert	Dubai	Creekside Park
	Tundra	Reykjavik	Laugardalur Park
	Undifferentiated highland	Quito	Parque La Carolina

creation of Englisher Garten in Munich, Birkenhead Park in Liverpool and Central Park in New York City. They were promoted on the basis of their potential contribution to public health and to social integration. The primary function of these nineteenth century parks was to provide walkways for strolling and roadways for carriage riding. In some cases former royal garden parks were modified to accommodate the circulation of pedestrians and carriages during the nineteenth century; however, a great number of nineteenth century urban parks were on urban sites that had not previously been used for parks. The creation of new parks provided the emerging profession of landscape architecture an opportunity to create arrangements of trees, walks, and avenues without any restrictions from existing garden trees. Shaded walks were seen as a important amenity in these nineteenth century parks as was the used of groves of trees to screen out views of adjacent city architecture and to create distinctive spaces within the parks. As a result, tree canopy cover in these nineteenth century parks is relatively high in comparison to parks derived from former royal gardens, but not as high as those large parks emanating from former hunting parks. Central Park in New York City is the classic American example of a nineteenth century park. Designed by Olmsted and Vaux in the mid-nineteenth century it has an area of 844 acres and supports over 150 species of trees (Barnard 2002). The overall tree canopy cover in Central is 65%, with some densely forested areas of the park reaching 100% canopy cover.

Large twentieth century parks have been built to accommodate active recreational pursuits. Soccer fields, softball fields, basketball courts, tennis courts, volleyball courts, climbing walls, skate boarding areas, swimming centers, ice skating rinks, and special areas for in-line skating are commonly included in these parks. Their inclusion reduces the space for trees. One example is Orange Park in Jacksonville. It is a close-fitted mosaic of four softball fields, two soccer fields, a swimming center, a skateboards park, a picnic area, and a natural area. Tree canopy cover in the activities portion of Orange Park is less than 1% and consists of about two dozen trees on the 80 acre site. A small unit of second growth longleaf pine/hardwood forest has been maintained in a corner of the park. This natural area covers only about 3% of the entire areas of the park, but has 100% canopy cover and supports seven native tree species.

Large cities built on hilly and mountainous terrain often contain areas that were too steep for urban development. These areas of steep terrain were left undeveloped but were eventually incorporated into an open space or park system. In some situations they served as grazing or agricultural land until their incorporation into the city. The steepness of the terrain of these units prevented their development for the park activities of interest (e.g., carriage riding, social walking, boating, athletic fields) in the nineteenth and early twentieth centuries. However, trails or roads were often built up the slopes to provide an overview of the city and a route for runners. Since the middle of the twentieth century some steep terrain remaining as open space in cities has been designated as parkland. The massifs in Rio de Janeiro are examples of terrain-determined parks that add to the visual character of a city and also serve a variety of more active recreational pursuits such as hiking and rock climbing. Pao de Acucar (Sugar Loaf) is famous for its massive granite dome that ascends above steep slopes leading down to the coastal plain. These steep slopes were formerly used to grow coffee, but were incorporated into the park system of Rio de Janeiro as the city grew around the mastiff.

In non-forested biomes the vegetation of terrain-determined parks is in many cases, following the removal of grazing livestock, succeeding back to the natural plant communities of the biome. Griffith Park in Los Angeles now supports a mosaic of oak woodland, chaparral, coastal sage scrub, and grassland communities; North Mountain Preserve in Phoenix is a mix of desert scrub communities.

Another twentieth century phenomena is the setting aside of natural areas and areas for nature restoration in large cities. For example, the Bukit Timali Nature Preserve in Singapore is a 405 acre nature preserve set aside to protect a tropical rain forest and to control torrents of water that develop during periods of intense rainfall. Large natural area parks have minimal infrastructure development and high percentages of tree canopy cover when they occur in forested biomes. Bukit Timali Nature Preserve has a single road leading to the top of the hill that is the highest elevation in the city (532 feet.). A few forest trails branch off this road. A visitor center and parking area have been developed at the lower elevation of the preserve. Tree canopy cover approaches 95 percent and tree species richness amounts to several hundred species.

Housing Areas

Tree are frequently planted around apartment building complexes and detached houses to provide shade, line sidewalks, and create privacy screening. The form of these plantings varies from the rigid rows of trees used in Moscow around apartment buildings to the naturalistic planting seen in suburban neighborhoods in Jacksonville. Two unusual designs for urban trees adjacent to housing areas were observed in Brasilia and Stockholm.

Brasilia

Brasilia was designed as a new capitol city for Brazil in 1957 by Lucio Costa. Among the innovative designs Costa created for Brasilia was the pattern of apartment buildings imbedded in an urban forest that was planted after the construction of the apartment buildings (Buchmann 2002). In Costa's plan two housing areas stretch out from a central axis of government and commercial buildings. Each of these housing areas is composed of quads of apartment buildings in a matrix of planted trees (Fig. 6.15). The trees form a more or less continuous canopy over the open spaces between the apartment buildings and the adjacent quads. Approximately 85% of the two housing areas are covered by tree canopies.

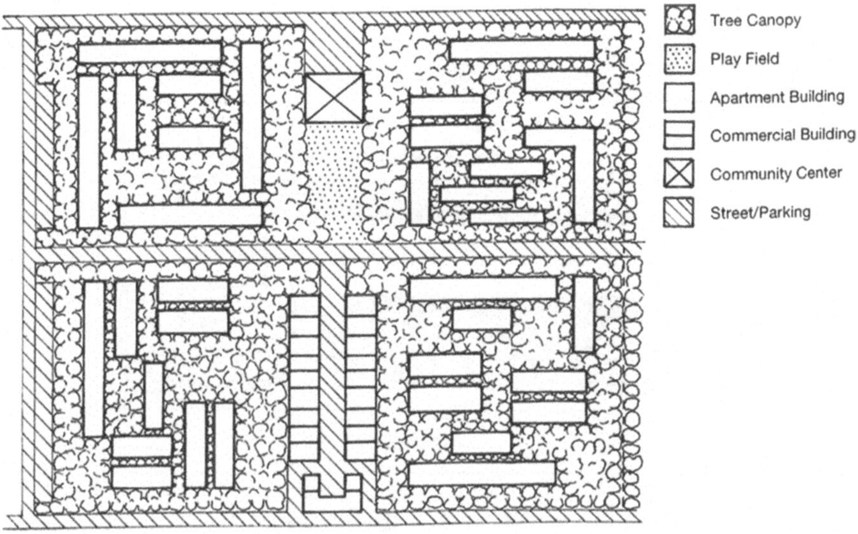

Fig. 6.15 Plan for housing quad, Brasilia

A variety of tropical species were used to create this continuous forest. Commonly observed species included aromo (*Acacia visco*), Argentine mesquite (*Prosopis alba*), ceibo (*Ceibo erythrina*), coral bean tree (*Erythrina crista-galli*), mango (*Mangifera indica*), royal poinciana (*Delonix regia*), silk tree (*Albizza julibrissin*), southern magnolia (*Magnolia grandiflora*), yellow flame (*Peltophorum pterocarpum*), and yellow trumpet flower (*Tecoma stans*). The trees had been planted in a semi-natural pattern that provided sufficient growing space for each tree. The canopies of the trees create a comfortable shade for the sidewalks through the forest matrix around the apartment buildings. Shade is important for outdoor comfort during the hot, dry season in Brasilia because of it proximity to the equator.

Stockholm

In the Nockeby, a suburb of Stockholm, a pattern of parallel strips of forest have been retained between rows of houses to provide a greenbelt. These forest strips were left when industrial forest land was converted into housing estates. They are approximately 100 feet wide and are flanked on each side by the backyards of detached houses (Fig. 6.16). The forest strips are residual areas of former commercial forest land at the edge of the city. As urbanization expanded into the commercial forests, strips of forest were incorporated into the developing suburbs. Trees of commercial value were generally removed from these strips prior to conversion of the commercial forest land to housing estates. The natural establishment of hardwood trees in these cut over strips has produced mixed stands of silver birch (*Betula*

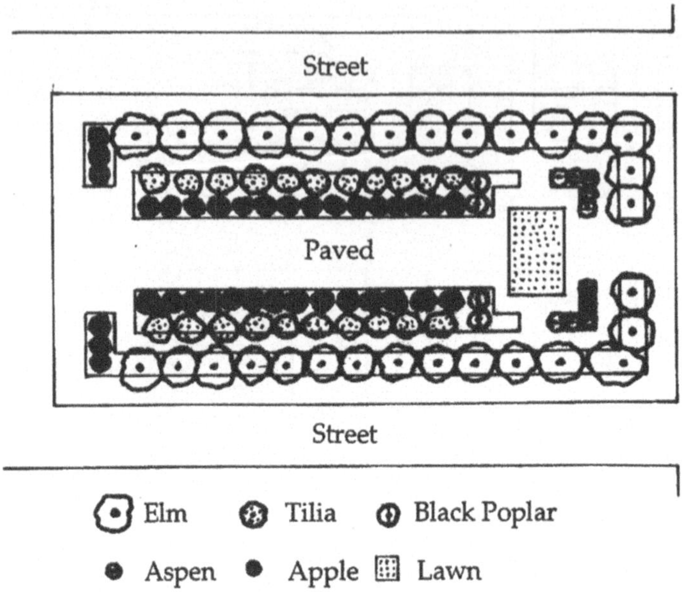

Fig. 6.16 Plan for small park with apple trees, Moscow

pendula), linden (*Tilia americana*), English oak (*Quercus robur*), rowan (*Sorbus aucuparia*), Norway spruce (*Picea abies*), and Scots pine (*Pinus sylvestris*). Unlike the forest matrix surrounding apartment buildings in Brasilia, these forest strips vary in tree density and species composition in relation to local site conditions and management history. Thinner soils on more rocky sites tend to support Scots pine while linden and English oak are found on deeper more productive soils. The presence of Norway spruce indicates areas where younger trees (too small to be of commercial value) of this species were growing at the time the commercial forest land was developed for housing. Most homeowners have gates opening into these forested strips to provide access for their children to play in the forest. Trails though these strips of forest have evolved over time as have campfire circles used by teenagers. The strips of forest also serve as windbreaks protecting adjacent houses. The presence of these forest strips in Stockholm suburbs enhances the quality of life and adds to the real estate values of the homes.

Urban Greenbelts

Urban greenbelts have been established at the margins of cities to define the limits of urban growth, provide space for outdoor recreation, and to improve air quality. In some cities agricultural land has been planted with trees to create forested greenbelts. The London greenbelt was proposed by Abercrombie in 1943–1944 as a web of open spaces flowing from the city center to the greenbelt (Turner 1991). It was composed of parks, roadside land, and agricultural areas. Its purpose was to control urban sprawl and provide for public access to rural parks and open space areas. The Abercrombie Plan has been replaced with subsequent plans that focused on park planning and the creation of a network of public pathways to connect the parks. A recent addition to these public pathways is London's "Green" bridge. A pedestrian and bicycle bridge built over a major thoroughfare two connect two parks formerly separated by the thoroughfare. The surface of the bridge has been covered with soil and planted with birch and rowan trees. Its width allows pedestrians to pass over the bridge and not lose the sense that they are still within a park.

Maintaining greenbelts for air pollution control dates back to the decision by the Roman Senate to prevent the development of agricultural lands associated with large villas at the margins of Rome (Cowell 1978). Strategically located forested greenbelts maintained around the city of Stuttgart, Germany in the twentieth century are noted for their value in the reduction of air pollution (Kaufman 1967).

Two of the cities studied for this book have forested greenbelts. Greenbelts have been designated in other cities in the survey, but these greenbelts are comprised of public parks and agricultural lands. They are not forested and were not chosen for study. The greenbelts surrounding portions of Beijing and Stockholm are comprised of forested land designed for the control of urban sprawl and the reduction in air pollution. These greenbelts were not designated for recreational purpose, but portions of them are being used informally by local citizens.

The greenbelt system in Beijing is comprised of tree plantations on former agricultural land that were established to define the limits of urban growth, reduce air pollution, and control soil erosion. The initial Beijing greenbelt was established in the 1970s outside of the fourth Ring Road. Subsequent urban growth has now surrounded much of this greenbelt and a new greenbelt is under construction beyond a fifth ring road that is also under construction. This fifth ring road and its adjacent green belt will define the final limit of urban growth for the city of Beijing.

Three greenbelts were examined in Beijing (Fig. 6.17). The characteristics of these three greenbelts are summarized in Table 6.6. Chinese white poplar was the primary species planted in the two greenbelts planted in the 1960s and Canadian poplar in the 1990s greenbelt. Every fifth or sixth row in the 1960s greenbelts were planted with either paulownia (*Paulownia tomentosa*) or Chinese ash (*Fraxinus chinensis*). Only Canadian poplar (*Populus x canadensis*) was used in the 1990s greenbelt. Uniform spacing between rows of trees was maintained at 10, 12, or 15 feet depending on the greenbelt. These same distances were used in spacing the trees along the rows. The rows were offset to increase the access of each tree to the resources of the site and to maximize wind reduction, and to screen dust and particulate matter from moving into the center of Beijing. All three greenbelts appeared to have from 95 to 100% survival. The trees had been pruned to a height of 18 feet.

The first greenbelt sampled had a footpath through it. Part of this greenbelt is used in the morning by nearby residents for tai chi exercises. A portion of the area adjacent to the entrance to an apartment complex was without any understory cover and showed evidence of compacted soil. Further from the entrance to the apartment

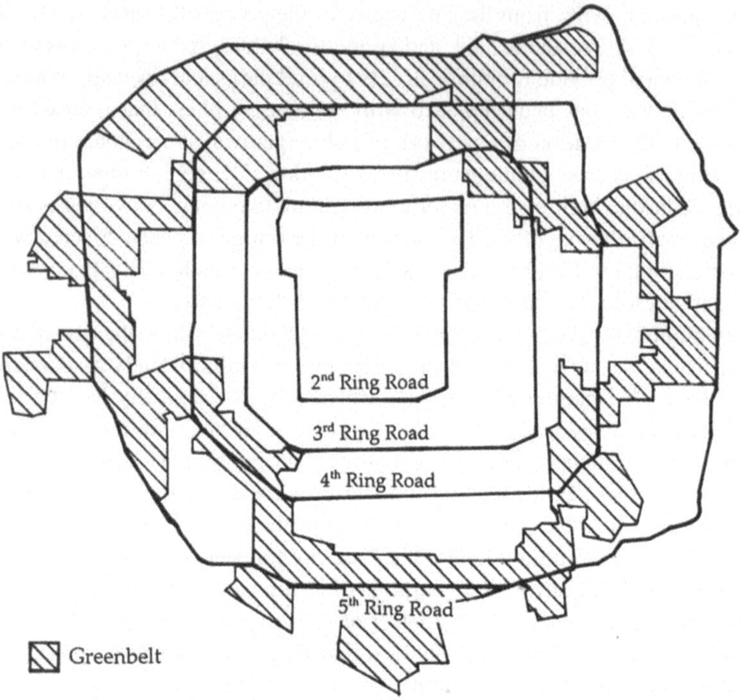

Fig. 6.17 Proposed Beijing greenbelts (1990 to 2010)

Table 6.6 Characteristics of greenbelts in Beijing

Characteristic	Greenbelt #1	Greenbelt #2	Greenbelt #3
Planting date	1960s	1960s	1990s
Primary species	Chinese White Poplar	Chinese White Poplar	Canadian poplar
	Populus tomentosa	*Populus tomentosa*	*Populus* x *canadensis*
Adjacent land use	Housing	Industrial	Agriculture
Secondary species	Chinese ash	Empress Tree	none
	Fraxinus chinensis	*Paulonia tomentosa*	
Number of rows	6	12	14
Spacing (ft):			
Between rows	15	12	10
Within rows	15	12	10
Average tree size[a]:			
DBH (in)	15	18	6
Height (ft)	48	70	50

[a]Of primary species

complex the understory had been invaded by a maple (*Acer tucutum*), sabina (*Juniperus sabina*), and Chinese ash (*Fraxinus chinensis*). Chinese white poplar (*Populus tomentosa*) seedlings and saplings were also established in the understory. The areas on either side of this greenbelt were being used for housing. Seed-producing trees of the three invading species were present. There was no evidence of maintenance in the greenbelt other than an early tree pruning. The windbreak and air pollution values of this greenbelt can be expected to deteriorate unless maintenance controls the succession of trees now evident in the understory. These invasive trees will eventually shade out adjacent poplar trees and disrupt the current wind barrier.

Interior portions of the second greenbelt sampled had been appropriated by homeless people who had cut down trees and constructed makeshift houses. This greenbelt was located in an area of large industrial buildings and was hidden from the sight by the building. This prevented authorities from observing the construction of the makeshift houses until they were occupied. In areas where the greenbelt had not been converted there was no evidence of invasion by other tree species. The dense canopy resulting from the close spacing of the trees in this greenbelt prevented the invasion by other species; however, the industrial use of the land adjacent to the second greenbelt did not support tree species other than Chinese white poplar. As with the first greenbelt there was no evidence of maintenance activities other than pruning.

The closely-planted third greenbelt sampled had not been invaded by other tree species. The ground surface beneath the tree canopy had a thick carpet of leaves. The greenbelt was adjacent to an agricultural area on one side and a canal on the other. No trees other than poplar were nearby that might have provided invasive seeds. The greenbelt showed no evidence of any human use.

The three tree greenbelts sampled are illustrative of the characteristics and uses observed at other greenbelts in Beijing. Invasion of these greenbelts by other tree species can be expected when seed sources are nearby. Human use will range from low impact recreation (walking and jogging, tai chi) to unauthorized housing. The newest greenbelts, due to their distance from housing areas and their visibility, can be expected to remain intact unless they are surrounded by development. More stringent management may be enforced in the final greenbelts planned for Beijing in order to maximize their value in limiting urban growth and air pollution reduction.

Greenbelts observed in Stockholm were either areas of industrial forest land adjacent to the margins of the city or parallel bands of trees that had been maintained as urban expansion spread into industrial forest land. Greenbelts on industrial forest land supported pure stands of either Norway spruce or Scots pine. Spacing in these stands was generally six by six feet at the time of planting. Thinning operations reduced this spacing to 18 by 18 or 24 by 24 feet. A major issue with these industrial forest greenbelts is their blockage of sunlight to adjacent houses and apartment buildings. Rydberg (1998) and Rydburg and Flack (1998) have proposed the creation of "step-shaped" forest edges to address this problem (Fig. 6.18).

Fig. 6.18 "Step-shaped" forest edge (Redrawn from Rydberg and Falck 1998)

Experimental application of their solution to the problem of forest greenbelts obstructing sunlight is underway in both Stockholm and Uema, Sweden.

Water's Edge

The water's edge provides a natural habitat for riparian woodlands. When riparian woodlands occur within urban areas they are considered special environments because of their habitat value, native tree cover, recreational value, and potential for environmental education. Urbanization often eliminates riparian woodlands by burying streams in various types of culverts or channels that eliminate space for trees. The loss of streams and their adjacent riparian vegetation has been documented in London by Barton (1962). He describes the fate of eleven rivers and streams that were buried or channelized during the history of the city. Although the fate of stream in many cities has often been like that of the rivers and streams in London, remaining water front areas formerly used for industrial purposes have been relinquished to be redesigned into commercial areas, parks, natural areas, and upscale housing. In London the Docks Project has converted portions of the old warehouses and shipyards along the Thames in to housing and public parks, in New York City old piers along the Hudson River have been modified for recreational uses, and in Buenos Aires a large section of riverfront land formerly controlled by the military is now a natural area. In these and other examples the planting of trees for aesthetic purposes and/or to recreate riparian habitat has been an important part of urban design.

The characteristics of water's edge environment determines the species of trees that can be used and the patterns of planting that are possible. The nature of the edge may or may not restrict the planting of trees. One normally observes a transition from a natural stream channel cross-section in the suburbs of a city to increasingly modified cross-section and channelization as one moves toward the center of a city. The first step in this transition is the resculpting of the channel cross-section from its natural form to a trapezoidal form. The trapezoidal form will accommodate greater volumes of water with increasing flows than natural cross-sectional forms and add an element of flood control to the stream. The next step in the transition is to harden the stream banks to prevent erosion. Various materials and designs have been used to accomplish erosion protection (Gray and Leiser 1982). Hardening of streambank often prevents their use as a site for tree planting; however, openings in the hardened surfaces can be designed to accommodate tree planting as can be seen along certain streams in Beijing. Channelization into box-shaped channels is usually the next step in the transition of urban streams. The concrete vertical wall of these engineered cross-sections do not provide any footings for tree planting. Treatment of the flood plain adjacent to the box-shaped channel may or may not allow for tree planting. When the adjacent area is also covered with concrete or other armoring material tree planting pits must be cut through the material to accommodate trees. This can also be the case for trapezoidal-shaped cross-sections with hardened walls and adjacent flood plain surfaces. The end point of the transition from a natural stream is complete burial of a stream in a box culvert. This ends any opportunity for the water's edge environment that the stream provides; however, rather ingenious designs have been developed in Tokyo to resurface a portion of the stream flow onto the top of a buried stream and create a simulated stream with some semblance of its riparian forest in the heart of the city (Ben-Joseph 1995).

The day lighting of streams in some cities has also provided an opportunity to expose buried streams and their water's edge environments for the planting of trees (Riley 1998). Larger rivers, lakes, and estuaries cannot be covered in the way smaller streams and creeks have been in urban areas. The banks of these bodies of water are usually converted to concrete or rock wall, thus eliminating any space for tree planting. Tree planting on the ground adjacent to these walls is, however, common.

Typically water's edge environments are planted with a single row of trees. In slightly less than one-half (54%) of the sites examined in the 33 cities, tree spacing was close enough to allow for a continuous tree canopy along the water's edge. Tree canopies over-arched streams or canals at only 26% of the sites. However, 34% of the sites had streams or canals narrow enough for crown canopies to arch over to contact the crowns of trees planted on the opposite side. Water's edge sites where tree canopy cover was continuous both along and across the water are often very attractive and very popular sites for strolling, resting, and enjoying the presence of a stream in the urban environment. At some sites two or more rows of trees were planted along the water's edge. In Paris, for example, Carolina poplar is planted along an artificially created first terrace of the Seine and London plane is planted along the edge of the former flood plain where it helps to define the river's edge and also to serve as a street tree. Both species were planted in circular tree pits cut in the paved surface of the first terrace and the flood plain.

In Moscow, portions of the edge of the Moscow river have been planted with two rows of trees on the top of the flood plane. A some sites silver birch and crack willow (*Salix fragilis*) have been planted with crack willow forming the row closest to the water. At other sites European ash (*Fraxinus excelsior*) and Norway maple (*Acer platanoides*) were observed with the ash in the row adjacent to the water. Trees observed along these portions of the Moscow river were planted in planting strips rather than the tree pits used in Paris.

A unique approach to the creation and recapturing of riparian woodlands was observed Tokyo were a design procedure referred to as "river raising" has been developed. Most of the riparian woodlands in Tokyo were eliminated during the last four centuries of urban development. This process began at the turn of the seventeenth century when Tokugawa, a local warlord who later became shogun, initiated large scale engineering projects that included the building of canals, channeling rivers, and landfill projects in and around the village of Edo (Cheng et al. 2000). This process of stream channelization and burial continued well into the twentieth century. By the 1970s hundreds of miles of streams had been channelized and buried in Tokyo. The Tokyo metropolitan area has a population of 11.8 million and an average population density 5400 people per square kilometer making it one of the worlds' most densely populated cities. The loss of riparian woodlands in Tokyo has meant a loss of contact with nature that is very important to the people of the city.

An interest on the part of urban planners to provide more green space in Tokyo has led to the adoption of the Green Master Plan in 1971. The main objective of this plan was to increase open spaces in the city to 20 m^2 per person by 1980. In the dense Tokyo urban environment the creation of open space was a particular challenge. Planners recognized urban streams as a network of open spaces that could be renovated to meet, in part, the open space objective of the Green Master Plan. Design solutions for incorporating the urban stream network into useable "green" open space were aimed at creating a hybrid between natural elements and stylized landscapes rather than the restoration of original stream channels and riparian vegetation (Ben-Joseph 1995). Inspiration came from traditional Japanese gardens and the water's edge compositions used in these gardens. Sensitive treatment of the water's edge is a tradition in Japanese garden design.

The creation of useable green space from the urban stream network had to recognize the constraints set by frequent typhoons and potential flooding. Any alteration of existing channelized streams had to accommodate the large runoffs common during the summer rainy season in Tokyo. Any new designs could not reduce the cross-sectional area of existing stream channels nor introduce trees into the channelized streams. Innovative design solutions ranged from the creation of simple access points on existing trapezoidal cross-sections to projects that involved the construction of artificial streams on top of existing channelized streams (Fig. 6.19).

Stream access projects were designed to provide physical access to restricted water channels in the bottom of existing channelized streams. Although many of these projects included reconstruction of a portion of the channel edge to make it appear more natural, they did not include the establishment of trees as a part of the restoration. Insertion of trees into existing channels would have required cutting through the concrete channel floor. This would create potential problems of water draining into adjacent underground structures. Placement of trees in channelized cross-sections would also increase the surface roughness of these channels and contribute to over bank flow and flooding of adjacent areas. Even without the addition of trees, the redesign of many stream channels resulted in relatively natural looking

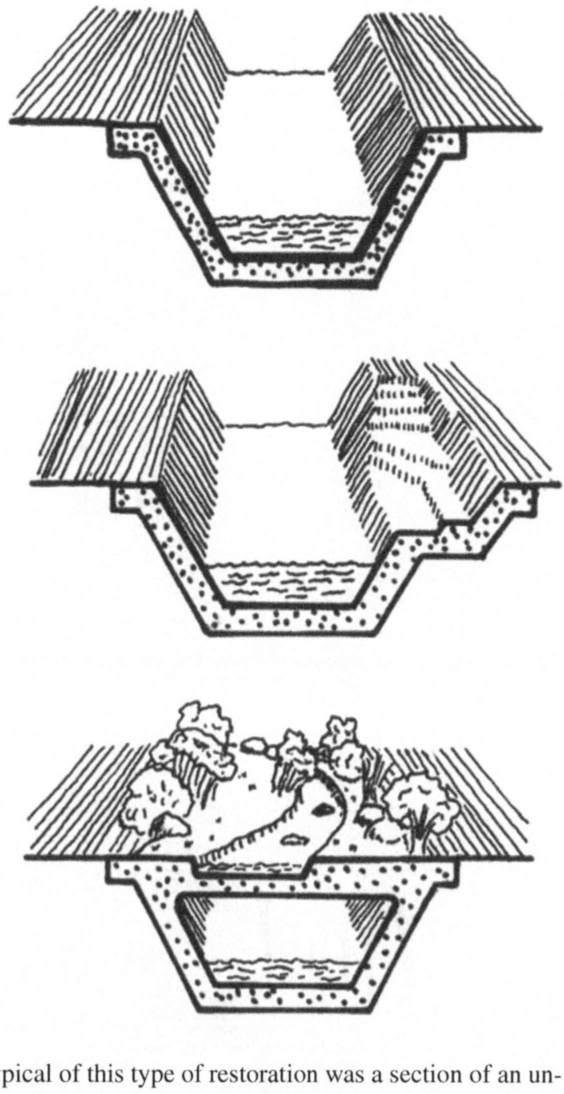

Fig. 6.19 Redesign of channelized streams, Tokyo (*top* = existing channelized stream; *middle* = access developed to reach stream; *lower* = "raised" stream)

streamside environments. Typical of this type of restoration was a section of an unnamed creek in the Shen Kawa district of Tokyo (Fig. 6.20). Here access to water in the concrete stream channel was made possible by widening the channel at several points to provide for staircases down to the water. A low water flow channel was designed to meander along the bottom of the concrete channel. The margins of this low flow channel are lined with rocks that help to protect a layer of soil on either side. The soil is held in place with nylon geo-fabric. Flood-resistance grasses have been planted on the soil along with a few broad-leaved semi-aquatic plants immediately adjacent to the low flow channel. Small islands have been inserted into the low flow channel. During the reconnaissance of the Tokyo urban forest in 1999 an ibis and a water turtle were observed using the low flow channel.

Fig. 6.20 Naturalized redesign of floor of channelized stream, Tokyo

Fig. 6.21 Raised stream, Tokyo

A more novel solution to the conversion of portions of the channelized urban stream network in Tokyo is known as "river raising". This involved covering an existing concrete channel and installing an artificial stream on the top (Fig. 6.21). Water depth in these raised streams can range from a few inches to as much as twenty inches in the deeper pools. Water is filtered and pumped to the upper level from which it flows down stream and back down to the lower channel at the end of the "raised stream" section. The design maintains the cross-sectional area of the concrete channel necessary to carry flood waters. The margins of the raised stream are landscaped using well-developed Japanese rooftop garden techniques. Trees are planted in special containers that are then covered with soil. Other plants are grown directly in the soil placed on top of the channel or planted in smaller versions of the tree containers. Footpaths, children's play areas and benches are incorporated into the design. One project constructed on a tributary to the Edo River is almost 3/4 of a mile-long and gives the appearance of a natural stream.

A channelized section of the Tama Gawa River has also been "raised" by covering the top of its concrete channel and constructing an artificial stream. The artificial stream channel is six feet wide and meanders in gentle curves over the fifty-foot wide top of the covered concrete channel for a distance of about 1/4 mile. Gingko *(Gingko biloba)*, black locust (*Robinia pseudoacacia)*, magnolias (kobusu) (*Magnolia kobus)*, and Japanese chestnut *(Castanea crenata*) trees have been used in the design of the forest along the "raised" stream. These trees have been kept to heights below 30 feet by top pruning in order to minimize their weight. The visual and social success of this design was apparent in its pleasing appearance and the large number of children and adults using the park along the raised stream as well as the raised stream.

These modifications of channelized sections of streams in Tokyo are applicable to channelized streams in other cities. The simple modification of the concrete stream channels offers a low cost way of improving the appearance and habitat quality of urban streams. The more expensive "raise stream" design provides an attractive means of incorporating more trees into a city.

Several cities that were studied were built on canals that transport people and materials throughout the cities. These canals represent a special category of the water's edge environment designed for transportation. Cities well-known for their canals are Venice and Amsterdam. Bangkok was also built on a system of canals, which still play an important function in transporting people and goods. An exploration of the designs used for tree planting along the canals in Bangkok illustrates the issues and values involved in the use of trees in the water's edge environment along canals.

The canals of Bangkok serve three important functions for the city: (1) as an aesthetic element in the urban landscape, (2) as a transportation network, and (3) as a system for moving storm water runoff. The canals of Bangkok add to the visual diversity of the city. They break up the repetition of streets and buildings. The boats that move along the canals provide a dynamic dimension to the urban landscape. They have a lively visual dimension because of the movement of water and boats. Bridges across the canals are places pedestrians often stop to watch boats and look

Fig. 6.22 Cross-section of Klong Lawd and adjacent streets, Bangkok

at the water, the backdrop of trees, and structures lining the canals. The canals also serve as a form of theater to celebrate Loy Kratong, a holiday to honor the rains that have watered the earth. Individuals launch tiny banana leaf boats, each carrying a single candle, in the canals. The Chao Praya River and lakes in city parks are also used to celebrate Loy Kratong, but the canals are the most accessible bodies of water for most of the people living in Bangkok to celebrate the occasion.

The canals in Bangkok can be divided into three categories on the basis of width. Large canals are 40 to 65 feet wide and are flanked by boulevards. Ample margins for tree planting and sidewalks have been left along the larger canals. The large canals form the backbone of the water transport infrastructure in Bangkok. They are the common routes for waterbuses and taxis. Water bus stops occur at regular intervals along the large canals where passengers assemble to board both waterbuses and taxis. Medium-size canals are 20 to 33 feet wide and form a secondary network connecting the larger canals. They also have space for trees along their margins. Streets parallel the medium-size canals, but occasionally one finds buildings flanking portions of the medium-sized canals. Water taxis use the medium-size canals, but these canals are too small for water buses. Small canals are ten to eighteen feet wide and seldom have margins available for tree planting. Sidewalks and buildings flank the sides of smaller canals. Many smaller canals have support beams that prevent their being used by boats. Smaller canals that do not have these beams are frequently used by the private boats of the people living along the canals.

Species composition and planting design is varied among the larger canals in Bangkok. A typical example of tree planting along the larger canals can be seen along Klong Lawd and the adjacent Thanon Atsadang boulevard (Fig. 6.22). Along most sections of Klong Lawd iron wood trees were initially planted at the end of the nineteenth century at spacing of 40 feet; however, tree mortality and bridge construction has resulted in the surviving trees often being 80 feet or more apart. Along that portion of the canal between Soi Sri and Thanon Ratchabopit, the spaces between the iron wood trees have been planted with jackfruit (*Artocarpus hetero-*

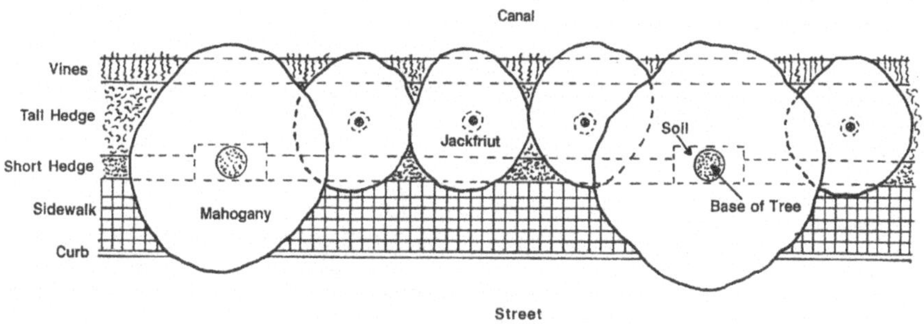

Fig. 6.23 Plan view of plantings along Klong Lawd and Thanon Atsang, Bankok

phylla) trees that are spaced 20 feet apart. At the base of the jackfruit trees, a hedge of Ixora (*Ixora stricta*) was planted to complete the design (Fig. 6.23). Another example of a planting design used along a large canal is illustrated in Fig. 6.24a. The cross section of the canal and its adjacent streets show planting both along the canal and along the sides of the street opposite the canal in contrast to plantings along Klong Lawd that are limited to the edge of the canal.

Traffic congestion in Bangkok has led to the widening of some avenues that parallel the canals. This is accomplished by removing the trees along the margins of the canals and cantilevering the new traffic lanes over the sides of the canal. A section of Klong Pohyom (Sathorn) has been modified in this way (Fig. 6.24b). The elimination of the trees from the sides of the canal has had a negative impact on the appearance of the canal. Furthermore, the shaded sidewalks formerly along the canal have been eliminated.

Medium-sized canals in Bangkok are often shaded by trees overarching them. This provides a very attractive environment for pedestrians and people moving along the canals in private boats or water taxis. Design features of these canals often included small docks or water taxi boarding areas. An example of a medium-sized canal with overarching trees is the Grand Palace Moat (Fig. 6.24c). Golden shower (*Cassia fistula*) and tamarind (*Tamarindus indicus*) have been planted along the Grand Palace Moat on 20 to 24 feet centers.

Another example of a medium-sized canal with overarching trees is Klong Bang Sue along Rama V Street. The canal separates the residential areas from the street. Angsana (*Pterocarpus indicus*) have been planted along the canal as well as the street on 20 to 24 feet centers. Tamarind (*Tamarindus indicus*), bo tree (*Ficus religiosa*), lagerstroemia (*Lagerstroemis calyculata*), and wattle (*Acacia auriculiformis*) were also found to have been planted on the banks of many canals.

Small canals in Bangkok vary in terms of space available for tree planting. Some small canals have strips 2–3 m wide along their margins where trees have been planted. Klong Wat Thethida, adjacent to Drok Sathien is a 5 m wide canal with queen's flower (*Lagerstroemia speciosa*) planted on 6–7 m centers along its margins (Fig. 6.24d). These trees form a closed canopy over the canal and along the edges of the canal. They provide a shaded passageway for both pedestrians and

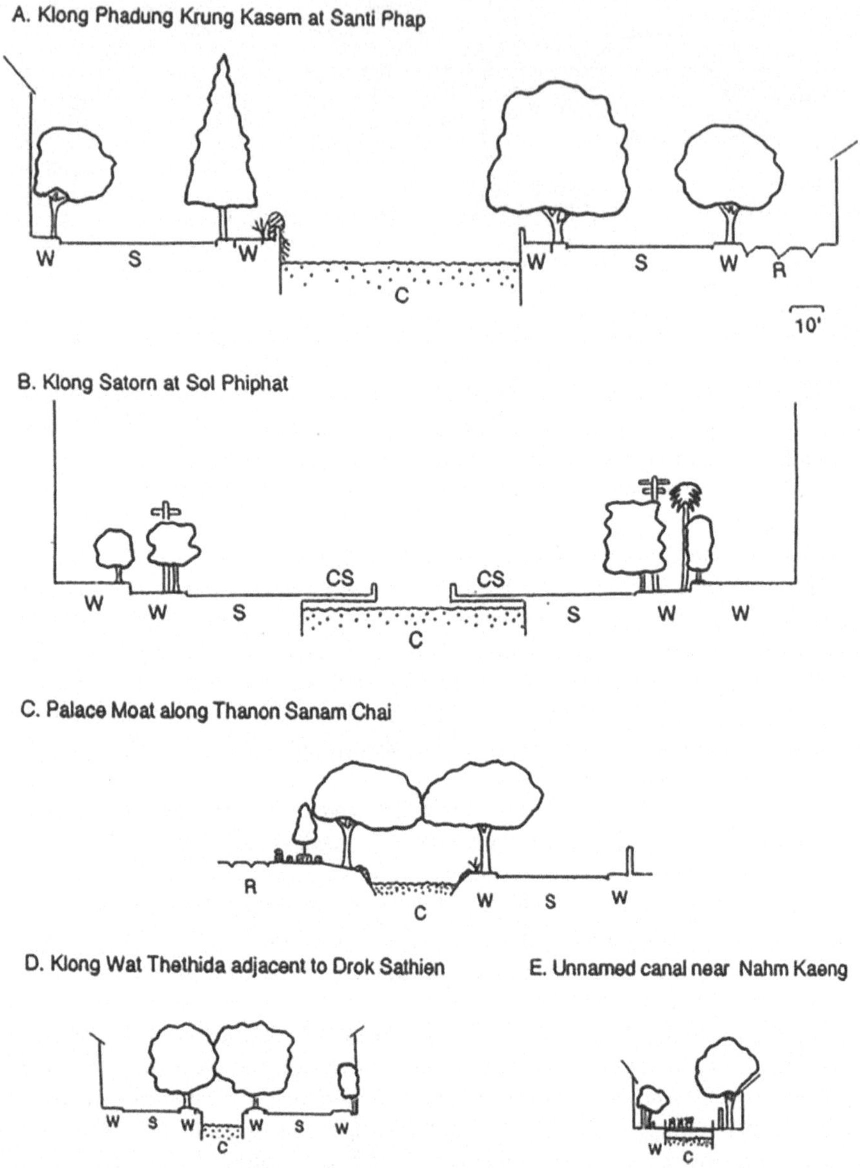

Fig. 6.24 Planting designs along major canals, Bankok

boaters. Other small canals do not have margins of sufficient width to allow for tree planting, but trees on adjacent private property arch out over these canals to provide shade to both the sidewalks and water surface. An example is Klong Wat Yannawa near Soi Rong Nam Kaeng (Fig. 6.24e). In addition to the trees planted along these small canals, the tops of the support beams (12 inch by 18 inch) across the canals were often covered with potted plants. These add to both the aesthetics of the canals and provide produce for home kitchens. Commonly planted species in these pots include lemon grass, peppers, ginger, and a variety of flowers.

Tree canopy cover in Bangkok has been estimated at 15%. Much of the canopy area is found in the many parks of the city and a large area of agricultural land supporting orchards that has been surrounded by recent urban development east of the Chao Praya River. The canopy areas of the trees along the canals in Bangkok amounts to something less than 1% of the city's total tree canopy cover. However, the value of these trees far exceeds their contribution as measured simply by canopy cover.

Cemeteries

The patterns of tree planting in cemeteries varies in different parts of the world. Trees are used in many cemeteries to reinforce entrances and delineate roadways and pedestrian paths. In some cases memorial tree groves are planted and individual trees planted adjacent to the graves. Allees of trees are in common use in American cemeteries while in other countries one finds single rows of trees delineating pathways. In cemeteries in Quito, for example, a single row of trees is planted down the middle of the major walkways through the cemeteries. These serve to orient visitors to the primary walk way through the cemeteries. Sections of cemeteries are often defined by rows of trees such as those used in the cemeteries in Tehran to delineate blocks of graves. Many of the cemeteries in the United States were derived from the rural cemetery concept of the mid-nineteenth century. The curving drives and paths used in these rural American cemeteries were not adopted in the design of late nineteenth century cemeteries in many parts of the world, with the exception of cemeteries in North America. Square and rectangular grid patterns for the arrangements of graves and sections of cemeteries were far more commonly observed in cemeteries outside of the United States. Although these patterns often included space for tree planting the resulting plantings were more formal and tended to avoid the woodland or forest appearance of cemeteries designed adhering to the rural design.

Considerable variation in the size of grave plots, the nature of the burial, and tree planting adjacent to graves occurred among the cemeteries studied. Larger individual grave plots were more common in cemeteries in cities in the United States and in the Russia. The layout of these cemeteries included tree-line roads and walkways. Typical Russian grave plots in both Moscow and Murmansk tended to be six by eight feet. The burial space itself was about two by six feet and was covered by cement or stone, often raised a few inches above the ground. In the surrounding space within the grave plot trees were planted. In contrast, cemeteries in Singapore and Sydney were packed with graves leaving no space for trees to be planted. In the Botany Cemetery in greater Sydney the individual burial site is two by six feet and often covered with cement or stone. No space exists between adjacent gravesites, the tombstones nearly touch one another. In a cemetery of this design trees planting

Table 6.7 Average tree canopy cover and range in canopy cover in cemeteries in different biomes

Biome	Canopy cover (%)	
	Average	Range
Tropical forest	40	1 to 80
Broadleaved evergreen forest	15	0 to 60
Deciduous forest	51	5 to 90
Coniferous forest	23	1 to 70
Mixed deciduous-coniferous forest	74	20 to 90
Mediterranean woodland scrub	24	10 to 50
Savanna thorn scrub	32	25 to 40
Grassland	15	5 to 40
Desert	2	1 to 2
Tundra	60	1 to 90
Undifferentiated highlands	1	<1 to 2
Range	na	<1 to 90
Average	34	na

was restricted to a short section of the entrance road (300 feet) that was lined with Canary Island date palm.

A unique aspect of cemeteries in Murmansk, Reykjavik, and Nuuk has been the tradition of planting trees on top of the buried coffin. One can often tell the oldest parts of the cemeteries by the heights of the trees which grow slowly in the tundra climate. Planting of trees on graves was believed to allow mourners to 'come in contact" with the departed by touching the tree, according to an Old Norse tradition (Valk 2006). This practice contributes to a greater tree density in these cemeteries and the highest cemetery tree canopy cover recorded (90%) in this study. Canopy cover ranged from less than 0% in individual cemeteries in Sydney, Tokyo, and Singapore to a high of 90% in Moscow, Murmansk, Reykjavik, and New York. Average tree canopy cover was 31.1% for all of the cemeteries studied. The average tree canopy cover in cemeteries in each of the biomes suggests a higher average cover in tree dominated biomes (Table 6.7). However, an examination of the range of cover values within the biomes reveals too much variation to support this conclusion. In the tropical forest biome the city of Singapore has a range in cemetery tree canopy cover from one to 80%. The Chinese cemeteries in the city are known for their lack of trees while the Moslem cemeteries are characterized by richness in tree species and abundant tree canopy cover. A somewhat similar spread in the percent canopy cover was observed between the new and the old cemeteries in the conifer forest biome city of Jacksonville where new cemeteries had less than 1% canopy cover and older cemeteries averaged 63% canopy cover. These new cemeteries conform to a pattern in the evolution of American cemeteries from tree dominated with upright tomb stones to treeless cemeteries with level tombstones designed to accommodate gasoline powered lawn mowers (Jackson and Vergara 1989).

Cemeteries in both the desert and tundra stand out in regard to canopy cover. The low average canopy cover and the narrow range in canopy cover in desert biome cemeteries is related to the need for irrigation to sustain trees in this biome. Although cemeteries in Dubai and Phoenix were equipped with irrigation systems, water was used conservatively in these cemeteries and tree densities were kept low. The high tree canopy cover in Murmansk and Reykjavik evolved from a need for wind protection in the harsh tundra climate. As with the planting of street and boulevard trees in these cities, tree density in the cemeteries was high. This high density resulted in the high percent canopy cover recorded as well as important wind protection for people visiting the cemeteries.

Freeways

The design aspects of moderns freeways, as far as tree planting is concerned, vary significantly from the parkways designed before the second world war. Traffic speed and the capacity to move large numbers of vehicles were design considerations in early parkways, but they did not prevent the incorporation of trees into median strips and along the margins of the parkways. The freeway system that was constructed in the United States following World War II was conceived of as a strategic system of limited access highways that could move materials and large numbers of vehicles during national emergencies. Patterned after the autobahn system developed in Germany prior to the Second World War, the American freeway system gave little thought to the tree planting that was the signature of the early American parkways. As freeways were built across American cities, the value of trees for noise reduction, dust control, and general aesthetics began to be recognized. Where space permitted trees were planted and became an important feature of many urban freeways.

Space is often limited when freeways are built through existing cities. More right-of way can usually be obtained in rural areas where the price of real estate is lower and political resistance to freeway construction is more limited. Along most freeways in urban areas the amount of space for tree planting varies with the width of the right-of-way and open space associated with off ramps and intersections. In the Los Angeles metropolitan area a network of 16 freeways have a combined length of about 350 miles. The adjacent right-of-ways average twenty feet in width and collectively amount to an area of about 1700 acres (Table 6.8). This area is more than 40% of the area of Griffith park, the largest park in Los Angeles.

Planting designs for freeways vary from a single row of trees along straight sections of freeways to more complex designs at off-ramps and intersections. The width of the right-of-ways available for planting also influences the designs used for tree planting. Planting space at freeway exits often provide opportunities for imaginative use of trees in landscaping. Seven different right-of-way patterns were observed along the six freeways studied in Los Angeles (Fig. 6.25). Tree planting on the right-of-ways was generally kept at a distance from the roadbed, but trees were planted very close to the roadbed where the planting surface and the roadbed were on different elevations.

Table 6.8 Width of planting space along Los Angeles Freeways

Freeway	Location	Left	Right	Average
Artesia	Downey Ave.	65	65	65
	Norwalk Blvd.	6	0	3
	Euclid St.	3	6	4
	Average	25	24	24
San Diego	Inglewood Ave.	15	15	15
	Washington Blvd.	30	30	30
	Getty Center Exit	30	30	30
	Mulholland Dr.	25	25	25
	Roscoe Blvd.	50	50	50
	Average	30	30	30
San Bernardino	Santa Ana Ave.	12	16	14
	Atlantic Blvd.	0	0	0
	Alameda St.	0	0	0
	Average	4	5	5
Santa Monica	Vermont Ave.	0	0	0
	Crenshaw Blvd.	20	30	25
	Overland Ave.	20	20	20
	Clover Field Blvd.	20	20	20
	Average	20	23	22
San Gabriel	Live Oak Ave.	25	25	25
	Peck Rd.	10	20	15
	Washington Blvd.	20	30	25
	Firestone Blvd.	30	30	30
	Del Amo Blvd.	10	10	10
	Average	19	23	21
Harbor	Slauson Ave.	6	10	8
	Rosecrans Ave.	0	0	0
	4th St.	20	10	15
	Average	9	7	8
Average				19

Much of the freeway right-of-way in the Los Angeles Metropolitan area supports trees that were planted for highway beautification, visual screening, noise reduction, and air pollution control. In Beijing broad right-of-ways adjacent to the freeways that circle Beijing have been used for the planting of forested greenbelts to define the limits of urban growth and to control dust pollution. Trees along Singapore's East Coast Parkway were planted to establish a grand boulevard from the international airport into the center of the city.

Sections of the right-of-ways along Beijing's Fifth Ring Road have been planted as a part of the greenbelt to define the limit of urban growth (Fig. 6.26). In these sections the freeway is flanked by five rows of Chinese poplar set 16 feet (5 m) apart. The trees have been planted on 16 feet (5 m) centers in each row. A center divider, which is planted with Chinese arborvitae, separates the opposing lanes of traffic. Trees planted in this fashion help reduce the noise of traffic and to control

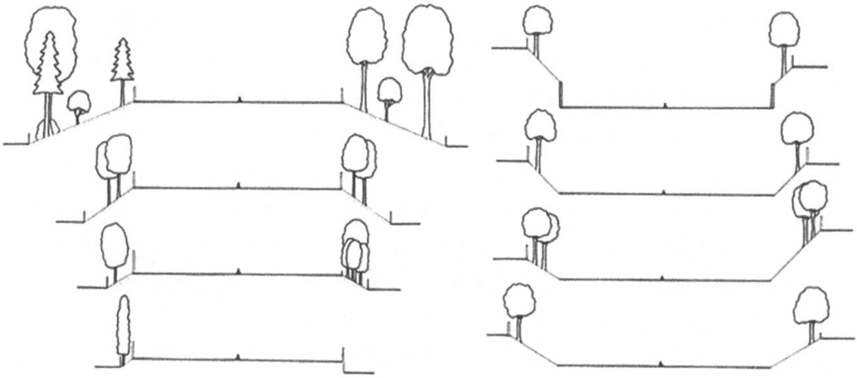

Fig. 6.25 Right-of-way planting patterns along freeways, Los Angeles

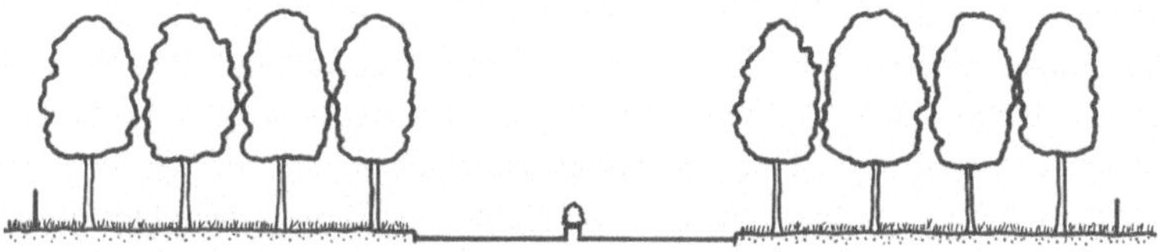

Fig. 6.26 Planting pattern along Fifth Ring Road, Beijing

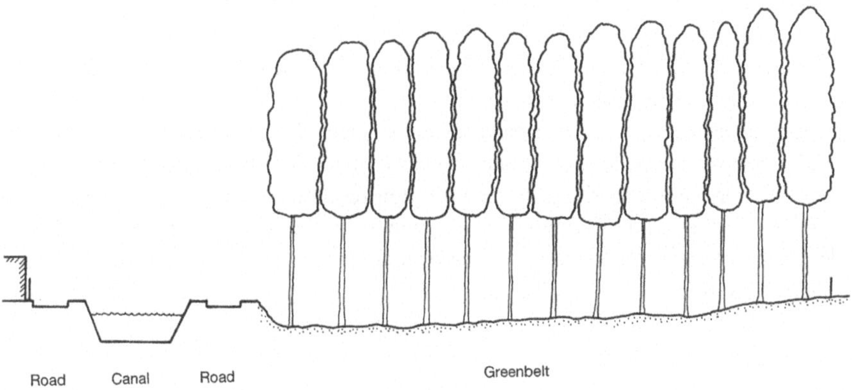

Fig. 6.27 Planting pattern along Fourth Ring Road, Beijing

dust from the freeway. An earlier planting design, that accommodates bicycle traffic, can be seen along sections of Beijing's Fourth Ring Road (Fig. 6.27). Here single rows of Japanese pagoda trees (*Sophora japonica*) and weeping willows (*Salix babylonica*) have been planted adjacent to bicycle lanes along each side of the freeway. The Japanese pagoda are on 16 feet (5 m) centers, while a spacing of 34 feet (10 m) was used for the willow. The Chinese arborvitae was also planted in the center divider of the Fourth Ring Road.

Two contrasting planting designs have been used along Singapore's East Coast Parkway. The straight portions of the parkway were designed to serve as auxiliary

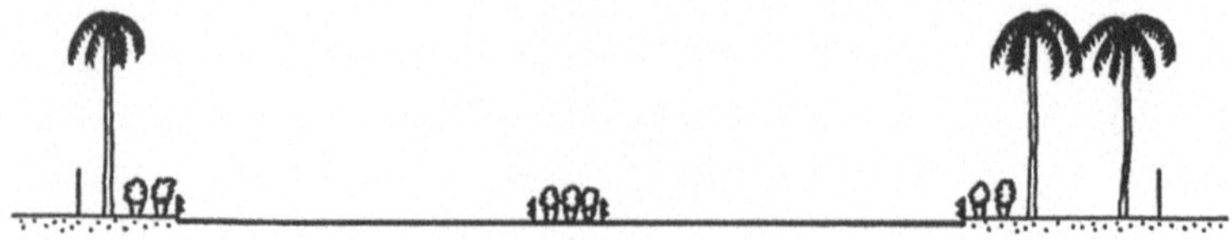

Fig. 6.28 Planting pattern along straight sections of East Coast Parkway, Singapore

Fig. 6.29 Planting pattern along curving sections East Coast Parkway, Singapore

landing strips in the event the Changi Airport was bombed. In these straight sections trees have been kept away from the margins of the freeway and the center divider supports low stature plants growing in containers that can be removed if the freeway is needed as a landing strip. Trees used along the right-of-ways of these straight sections are maintained at a distance of about 65 feet (20 m) from the edges of the road and serve to screen the adjacent housing areas, golf courses, and the East Coast Park (Fig. 6.28). In the curving sections of the East Coast Parkway the center divider is planted with trees as are the margins of the freeway (Fig. 6.29). The trees are planted on 50 feet (15 m) center both in the center divider and in planting strips that separate the freeway from adjacent streets. The branches of these rain trees reach out to almost touch over the roadways.

Many freeway right-of-ways in urban areas have a low tree density or no trees at all. The freeway environment often has space to trees. Additional tree planting along urban freeways can contribute to air pollution reduction, reduce peak runoff from the right-of-ways, and often improve the view from the road. The freeway right-of-way is one of the more disregarded environments for urban forest development in the world's cities.

Rooftop Gardens

Designs used in rooftop gardens that support trees varied considerably. No particular pattern in the arrangements of trees on rooftop gardens were common to the gardens rooftop gardens survey for this study. Public rooftop gardens used fewer trees than private gardens. In the public gardens the trees were often placed so as not to block views from the rooftops nor to obstruct views of flowerbeds or other decorative rooftop features. Many public rooftop gardens in New York do not have trees (Fig. 6.30). Not only more trees and more tree species were observed in private rooftop gardens. Trees were often strategically located in these gardens to provide wind protection and to ensure privacy (Fig. 6.31). The tallest trees observe in

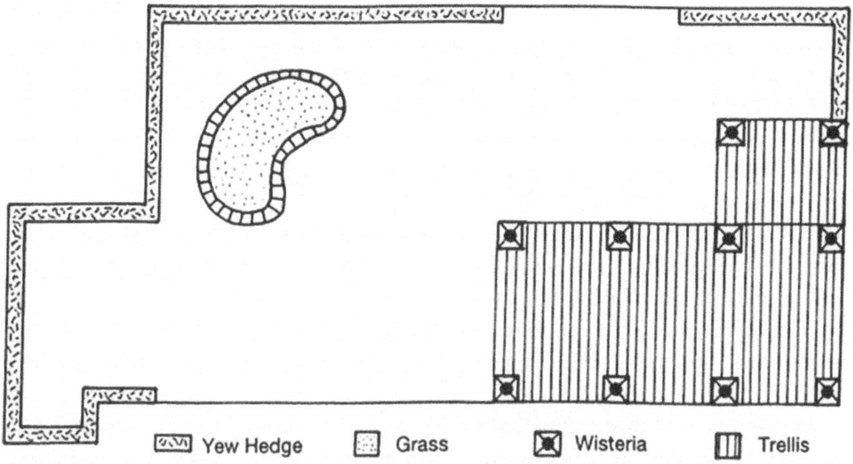

Fig. 6.30 Metropolitan Museum of Art roof garden, New York City

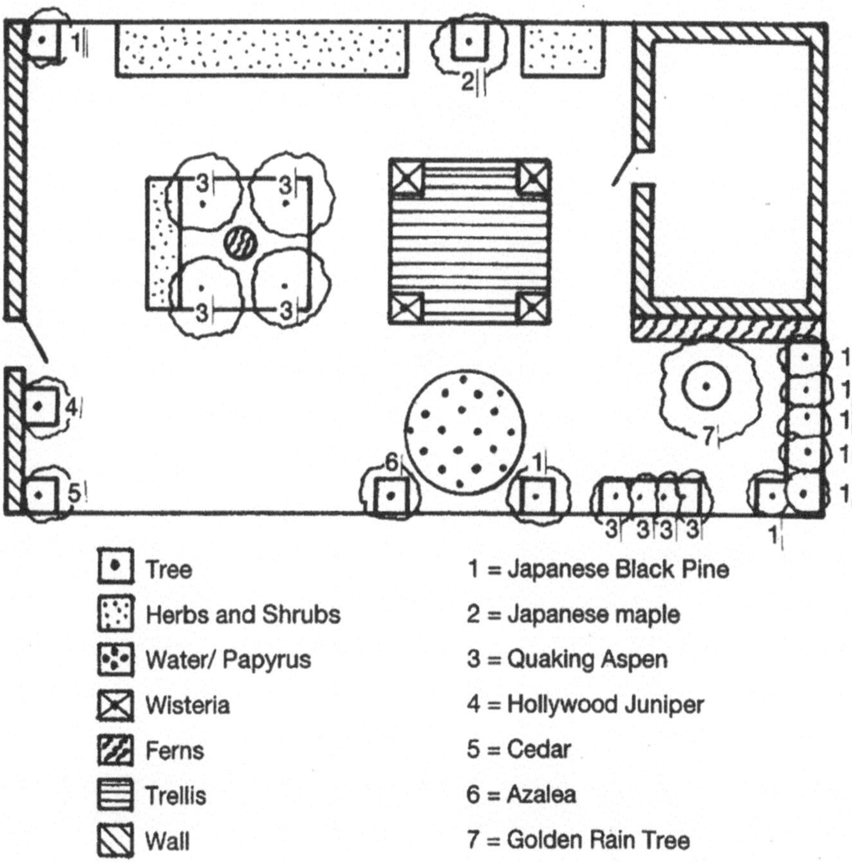

Fig. 6.31 Private roof garden, New York City

rooftop gardens in New York were under 15 feet in height. They were kept low in stature because of the high wind velocities on rooftops. Tree canopy cover on rooftop gardens observed in New York, Singapore, and Tokyo averaged less than 10%.

Certain engineering obstacles must be overcome if trees are to be installed on the roofs of buildings. Trees, the soil required for their growth, the watered stored in the soil, and tree containers are heavy and require considerations beyond those necessary for groundcovers, grass, and shrubs. Many roofs were not designed for the additional weight associated with trees. Trees may, however, be accommodated on these roofs by placing them on platforms to distribute their weight over large surface area. Because of their height, trees develop wind resistance that results in downward pressure during periods of high wind velocity. This pressure may rupture the waterproofing membrane that protects the roof if there is not a sufficiently thick protective layer between the tree container and the waterproofing membrane. The higher wind velocities encountered on the roofs of buildings require that trees in roof gardens be anchored. Anchoring is usually achieved by cables attached to the roof or by staking. A discussion of these design variables in presented in Osmundson's book on rooftop gardens (Osmundson 1999).

Chapter 7
Function

In the 1980s Rowan Rowntree (1984, 1986, 1988) edited three special issues of Urban Ecology and Landscape and Urban Planning on urban forestry in which he stressed the functional roles of trees in cites. These functions can be categorized as ecological, economic, aesthetic, and psychological. Trees occurring in several urban environments simultaneously serve several of these functions. London plane trees contribute to the aesthetic design of many avenues and boulevards while reducing urban heat islands and air pollution. Evergreens, such as Italian cypress, provide psychological support for mourners visiting cemeteries as well as contributing to the aesthetic characteristics of burial grounds. In some cities the functional role of some trees stands out due to special plantings to address particular functional roles. These include the use of trees for air pollution reduction, reduction of landslide potential, fruit production within cities, forest products, and storm water management.

Use of Trees for Air Pollution Reduction

Trees reduce air pollution by absorbing gases though their stomata and the deposition of particulate matter on their leaves and branches. The role that urban forests play in air pollution reduction has been made evident by the research of David Nowak of the U.S. Forest Service. Nowak pioneered the development of computer models in the early 1990s to calculate air pollution reduction by urban forests (Nowak 1994a, b). The "Urban Forest Effects Model" developed by Nowak and Crane (2000) subsequently was incorporated in the i-Tree tools for assessing and managing forests and community trees (https://www.itreetools.org/about.php).

The "Urban Forest Effects Model" has been used to calculate air pollution reduction in two cities surveyed for this book: New York and Beijing. These studies demonstrate the air pollution reduction value of trees in urban area.

J.R. McBride, *The World's Urban Forests*, Future City 8,
DOI 10.1007/978-3-319-52108-4_7

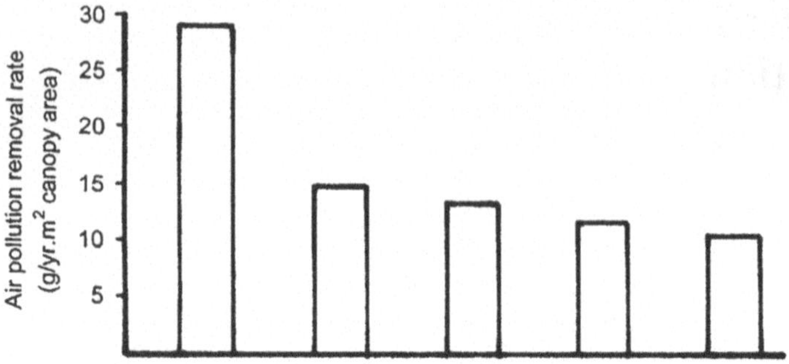

Fig. 7.1 Air pollution removal by trees in Beijing and selected American cities

Nowak and Crane (2006) estimated that trees in New York City removed an estimated 1821 metric tons of air pollution at an estimated value to society of $9.5 million in 1994. Air pollution removal by the New York urban forest was greater than that in Atlanta (1196 ton/year; $6.5 million) and Baltimore (499 tons/year; $2.7 million). Urban forest mediated air quality improvement in New York City during the daytime in the spring and summer averaged 0.47% for particulate matter, 0.45% for ozone, 0.43% for sulfur dioxide, and 0.30% for nitrogen dioxide.

Yang et al. (2005) used the "Urban Forest Effects Model" to calculate air pollution removal by the Beijing urban forest. Their findings showed a removal of 1261 metric tones of air pollution in 2002. The annual removal of ozone amounted to 256 metric tons, nitrogen dioxide 132 tons, sulfur dioxide 101 tons, and 772 tons of particulate matter (PM10). Beijing's urban forest removes more air pollution that the urban forests of New York City and other American cities when air pollution reduction is standardized on the basis of air pollution removal per square meter of canopy cover (Fig. 7.1). The higher removal rate in Beijing is a function of the higher concentration of air pollutants in Beijing.

Trees also emit volatile organic compounds that can contribute to ozone formation when the concentration of nitrogen oxides in the atmosphere is sufficiently high. These conditions exist over urban areas where the exhaust from vehicles increases levels of atmospheric nitrogen oxides. The contribution of urban forests to the formation of ozone is difficult to calculate because volatile organic emissions from trees are temperature dependent and emission rates vary among tree species. Extensive urban forest canopy cover can lower urban air temperature and reduce the emissions of volatile organic compounds. Another variable in calculating the potential amount of ozone produced over cities due to the emission of volatile organic compounds from trees is the spatial and temporal concentration of nitrogen oxides over a city. Urban air pollution specialists have tended to estimate the amounts of volatile organic compounds released from trees as a measure of the potential for ozone formation (Cardelino and Chameides 1990; Geron et al. 1994; Taha 1996; Nowak et al. 2000).

Trees in Beijing emitted 202 tons of isoprene and 36 tons of monoterpene with a total ozone forming potential of 1953 tons in 2002 (Yang et al. 2005). Emission of

Table 7.1 Air pollution reduction under three urban forest planning alternative in Ningbo and Hefei, China

	Air pollutants removed (T/year km^2)							
	O^3		NO_2		SO_2		TSP[a]	
Urban forest alternatives	Ningbo	Hefei	Ningbo	Hefei	Ningbo	Hefei	Ningbo	Hefei
Existing species composition and canopy cover	0.52	1.36	0.16	0.67	0.13	0.90	1.59	4.10
Existing species composition; 30% canopy cover	1.61	1.61	0.46	0.79	0.37	1.06	4.81	4.80
Improved[b] species composition; 30% canopy cover	1.86	1.65	0.60	0.80	0.43	1.08	6.81	4.82

[a]Total suspended particles
[b]Species with high capacity to remove air pollutants
McBride et al. (2001a, b)

these compounds in New York City are estimated to be 394 tons/year of isoprene and 93 tons/years of monoterpenes (Nowak 2005). An estimated 3993 tons of ozone could be formed from these compounds in the air over New York City. These potential numbers are much higher than the actual ozone formation because of the limited amount of nitrogen dioxide commonly found over urban areas (Benjamin and Winer 1998). However, they show a potential problem that could be caused by trees in Beijing and New York City. The nitrogen dioxide emission in Beijing is increasing rapidly because of the increase in the number of automobiles (Liu and Ren 2003). Volatile organic compounds emitted by trees may contribute to the formation of ozone in the future because of the high photochemical reactivity of these compounds in the hot summers, especially in the slow air mass movement characteristic of Beijing (Chameides et al. 1988).

The role urban forests play in air pollution reduction was demonstrated in a comparison of alternative urban forest planning scenarios in two Chinese cities (McBride et al. 2001a, b). In this study the "Urban Forest Effects Model" estimated the air pollution reduction value of the three different urban forest planning alternatives for the expansion of each city. Both cities are undergoing rapid urban growth and are planning to expand their area by 70% over a period of 15 years. The alternatives considered for expanding the urban forest of each city were (1) maintain the same species composition and tree canopy cover as occurs in the existing city, (2) maintain the same species composition, but increase the tree canopy cover to 35%, and (3) use tree species that are highly ranked for their air pollution reduction capacity and increase the tree canopy cover to 35%. The air pollution reduction value of these three planning alternatives is shown in Table 7.1. In both cities the greatest air pollution reduction was achieved by alternative three. The model indicated that the percentage of air pollution reduction/square kilometer would be greater in the city of Ningbo than in Hefei. This was due to the lower tree canopy cover that exists in Ningbo (10%) in comparison to Hefei (25%). Where existing urban forest canopy cover is low a greater increase in air pollution reduction can be achieved by increasing canopy cover as opposed to changing species composition.

Urban forests have not reached their potential for air pollution removal in the world's cities. That potential can be reached through increasing the tree canopy cover in most cities and adjusting the mix of tree species to maximize pollution reduction while minimizing the omission of volatile organic compounds.

Landslide Stabilization

Landslides in the United States are estimated to cause about $1.5 billion in property damage annually. Much of this damage occurs in urban and suburban areas. The Loma Prieta earthquake near San Francisco, California in 1989 triggered thousands of landslides throughout a 5400 square mile area. An estimated $10 million of damage to houses, other structures, and utilities was caused by these landslides. Renewed interest in the establishment of tree cover on many slopes, subject to land sliding in the San Francisco Bay Area, occurred following the Loma Prieta earthquake. Two years after the 1995 earthquake in Kobe, Japan a community tree planting program was launched to stabilize hillsides around the city.

Establishment of forest plantations and the designation of protection forests in mountainous topography around villages and towns to reduce the risk of landslides and avalanches is an old practice. As early as 1342 a community in Switzerland established a protection forest to minimize the hazard of avalanches. Between 1535 and 1777, 320 protection forests were designated by communities in Switzerland (Kittredge 1948). Techniques for tree planting to minimize the risk of landslides, soil slips, and other forms of mass wasting have been handed down over the centuries. These techniques have been reviewed and put in the context of modern engineering solutions by Gray and Leiser (1982).

Quito, a city built in the Andes and surrounded by steep mountain slopes, has major landslide problems. An analysis of newspaper reports by de Norri et al. (1986) suggest there were 67 landslides causing a negative impact to people and property between 1900 and 1986. These landslides were often coupled with mudflows. Thirty-eight mudflows were reported in newspapers from 1900 to 1986. The Quito Municipal Hydrology Department estimated that 300 landslides occurred in and around the margins of Quito from 1910 to 1990, the great majority of which did not result in property damage (Murray 1999). Typical of the problem was a Quito landslide in April of 2000 that killed 15 people, forced the evacuation of 180 others, and destroyed 30 homes (Associated Press 2000). The seriousness of the problem prompted the Inter-American Development Bank to approve a US$ 40 million loan to the Water and Sanitation Department of Quito in 2002 to support the first phase of a US$ 110 million program to expand water and sanitation services and reduce flooding and landslides. Management and expansion of forest plantations to reduce landslide risk were a part of this program.

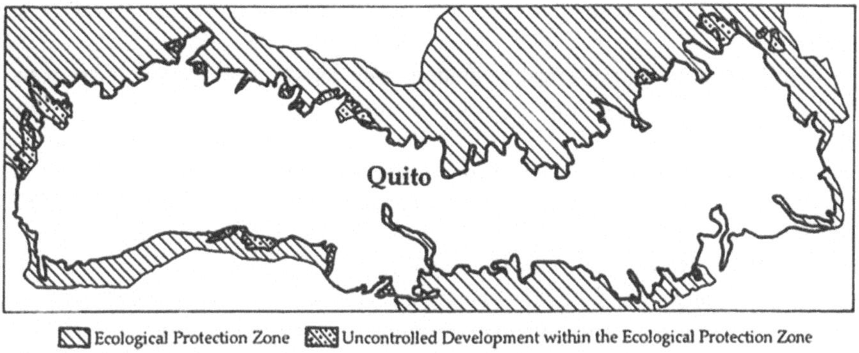

Fig. 7.2 Ecological protection zone and areas of uncontrolled development, Quito

Forest plantations had been established in the Andes following the introduction of eucalyptus and Monterey pine in the nineteenth century (Moore et al. 1984). These early plantations were planted to reforest the highland zone and to provide forest products (Chapela et al. 2001). The establishment of forest plantations for landslide protection began around Quito in the 1960s on steep slopes with unstable soils. Prior to the 1960s much of the foothills flanking the city had been used for agricultural purposes. Protective forest planting was employed as a means of reducing the risk of landslides as urban development spread to the base of the foothills. The total area of forest plantation in the foothills adjacent to Quito was 6889 acres in the early 1990s (IMQ 1992).

Murray (1999) measured the area of eucalyptus shown in aerial photographs in a 2.8 million hectare area in a northwestern portion of the foothills adjacent to Quito. The results of these measurements indicate a period of tree planting from the 1950s to the early 1970s. Planting during this period resulted in 40% of the area being covered by mature eucalyptus plantations in 1976. Aerial photos taken after 1980 show a decline down to only 8% of the area supporting eucalyptus by 1995. This decline was primarily due to illegal urbanization of the northwestern foothill area. In 1956 only 0.4% of Murray's study area was classified as urban development, by 1995 72% of the area was urban. Tree cutting and the clearing of ruderal vegetation is used as a "signal" of private control over land without legal ownership on the part of the individual clearing the land in many portions of the Andes. It has been a common visible signal of both land possession by urban squatter and other examples of illegal subdivision (Murray 1999).

In 1980 an ecological protection zone was established by the city to protect the existing forest plantations and to control unauthorized development of property in the Pichincha foothills along the western edge of the city (Fig. 7.2). The area was given federal protection in 1983 when it was designated as "Protected Forest and Vegetation". This protected area was named the "Bosque Protector Pichincha" and had an area of 18,000 acres. A report prepared at the time found only 18.7% (3857 acres) of the protected area supported forest plantations. Much of the area (57%)

was either enclosed pasture or open grassland known as paromos (Moore et al. 1984).

Two slope stabilization forest plantations were surveyed in Quito as part of the study of the world's urban forests. The first was a blue gum (*Eucalyptus globulus*) plantation on the slopes of Metropolitan Park and the second was a plantation of Monterey pine (*Pinus radiata*) planted on the slopes above La Cima de la Libertad. Metropolitan Park is within the federally declared "Bosque Protector Pichincha". The Monterey pine plantation occurs in the eastern foothills above the city and is not part of the "Bosque Protector Pichincha".

The eucalyptus plantation was planted on a two by two meter (six by six feet) spacing, although some variation in the spacing existed because of slope conditions. Trees in this plantation form a continuous canopy covering 100% of the ground surface. The existing trees were sprouts eight to twelve inches in diameter and fifty to sixty feet in height. Eucalyptus sprouts of this size are ten to fifteen years old when they occur on productive soils. These sprouts were the result of an earlier harvest operation. It appeared that coppicing, periodic cutting or pruning to encourage suckering had been done in some portions of the stand on more than one occasion. The coppicing was central to the management of the eucalyptus plantations established for landslide control. Periodic coppicing of eucalyptus will maximize leaf area near the ground and minimize runoff and erosion over the long run. A thick layer of leaves and eucalyptus bark covered the surface of the ground. No gullying or surface erosion was in evidence. There was no evidence of cattle grazing in this plantation.

The Monterey pine plantation was planted in contour rows twelve feet (3.5 m) apart with a spacing of six to eight feet between the trees. Trees averaged eighteen inches in diameter and 85 feet in height. Some trees had been removed from the plantation, but not in any systematic fashion. Their removal did not appear to be part of any organized management of this pine plantation. The age of recently cut stumps indicated this stand was 35 years old. Wind throw resulted in the opening of the crown canopy of the plantation. Overall tree canopy cover amounted to 85%. Livestock had access into the plantation through a broken fence. The ground layer supported a thick layer of pine foliage and a large crop of mushrooms. Contour drainage ditches had been dug at intervals of about 50 feet to carry runoff water into a down slope drainage channel. Although the plantation was young, it had the appearance of a stand that was beginning to break apart. Monterey pine is a short-lived species that tends to suffer from wind breakage and wind throw on exposed sites once it grows to heights above 50 feet (McBride and Froelich 1984). Unlike eucalyptus it cannot be managed on a coppice basis to renew stands before they begin to fall apart.

Another aspect of the forest plantations on the slopes above Quito is the unauthorized development of spontaneous human settlements in or adjacent to these plantations. Uncontrolled development of land has been a phenomenon around many large cities in South America (Violich 1985). A proliferation of spontaneous housing began on the slopes above Quito in the 1970s. Concern has developed over the fire hazard associated with many of these developments. A special forest fire safety program has been developed by the Quito Municipal Forestry Department to educate people living in and around the margins of forest plantations. Community leader in these developments have embraced and encouraged these programs. Fires in these plantations would have a direct effect on the people living there and contribute to the landslide problems facing the people living further down the slopes.

Urban Orchards

Tree planting in and around the margins of cities for the production of fruit and nut crops is an old tradition dating back to the use of date palms in ancient Egypt. The use of such trees may have predated the introduction of trees for shade or aesthetic purposes in cities. In the poorest shanty towns of New Delhi, Soweto, and the hutongs of Beijing one sees fruit and nut trees planted in the smallest of spaces. They are testaments to the faith in a future crop and the human desire to improve one's habitat.

Productive trees have also been designed into the fabric of cities to provide fruit and nuts for the municipal coffers or the common citizen. Charlemagne's court issued regulations for the administration of cities within the Holy Roman Empire in the eighth century AD (*Capitulare de villis vel curtis impreii,* 795 AD). These regulations stated that apples, pears, chestnut, peaches, quince, hazelnut, almond, mulberry, bay, stone pine, fig, walnut, and cherries were to be grown by the city for the public good (Fischer-Benzon 1894). New Islamic cities built in Spain in the twelfth century AD, such as Elche, were surrounded with date orchards to improve the nutrition of citizens and serve the economy of these cities.

Space for urban gardening and the maintenance of fruit and nut trees declined during the Industrial Revolution in northern Europe in the late eighteenth and early nineteenth centuries. The construction of blocks of tenement housing eliminated the pre-existing private garden space that had been used for kitchen gardens and the few fruit and nut trees growing in them. Many of the workers recruited for the new industries were brought into the cites from rural areas. They came with a tradition of vegetable gardening and in some cases knowledge of the care of orchard trees. The concept of providing space in towns and cities for the gardening needs of the larger community began in England in the eighteenth century (Warner 1987). Vacant urban land was acquired and made available to factory workers to supplement their diets and to provide some relief from the dismal nature of the tenement districts. Similar urban gardening programs were initiated by the Swedish National Road administration in Stockholm in the nineteenth century. In Moscow an extensive system of community gardening areas was established around the margins of Moscow during the periods of industrialization before and after World War II.

A community gardening movement evolved in many cities the United States in the 1970s and 1980s (Fox et al. 1985; Francis et al. 1994; Lawson 1997). Unlike the victory gardens established in parks and other open space areas in the United States primary for food production during the First and Second World Wars, these

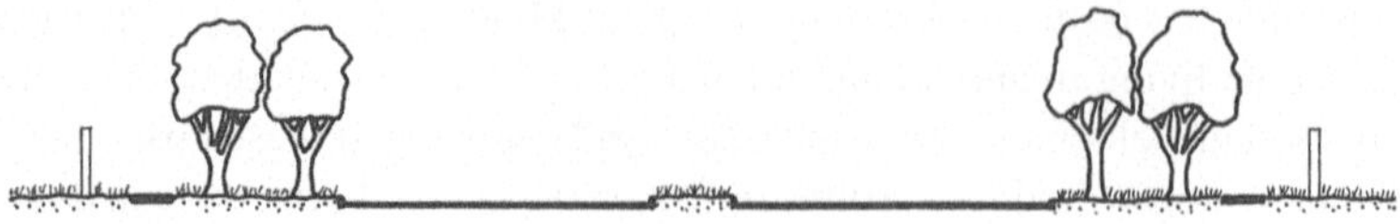

Fig. 7.3 Apple trees planted along boulevard, Moscow

community gardens were stimulated by a number of factors including the desire for improvement of the neighborhood, social symbolism, economics, community building, and youth training (Lawson 1997). Although initially concerned with the production of vegetables for urban gardeners, many community gardens provided space for berry and tree crops. Other community gardens have seen an evolution from subsistence gardening that depended on the production of vegetables to recreational gardening where flowers and fruit trees now occupy a fair portion of each gardeners allotment.

The value of tree fruit and nut crops in urban areas can be substantial. Kollin (1991) estimated that the annual production of fruits and nuts by trees in private and community gardens in San Jose, California, a city of 950,000 had a value of $10,000,000. An examination of community and private gardens occurring in the 33 cities visited in this study of the world's urban forests revealed the common use of apple, cherry, and plum trees in temperate forest and grassland biomes, mangoes, papaya, breadfruit, and bananas in the tropical forest biome, and date palms and citrus trees in desert biomes.

In addition to community gardens where fruit and nut trees are grown, some cities have allocated public space for orchards. Contrasting use of public open space for orchard planting was observed in Moscow and Tehran. The approaches used in these two cities show how the incorporation of orchards into the urban forest can serve to support community finances and provide food for urban populations.

Apple trees are widely planted in Moscow to provide apples for local citizens. Apple trees occur as orchards adjacent to apartment complexes, in both large and small urban parks, and along boulevards. Men in business suits can be seen picking individual apples to eat at work and older women picking sufficient quantities for cooking pies and other apple treats. They carry special hooked poles for bending down the branches of the apple trees. Children are also important consumers of these public apples.

Sections of the boulevards adjacent to the University of Moscow campus have been planted with apple trees. The trees were in two rows in the parkways along the boulevards (Fig. 7.3). In these broad parkways the trees had been set back 15 feet from the edge of street surface to avoid apples from falling onto vehicles. A spacing of 15 feet was maintained between the two rows of trees and between the individual trees in each row. A few apple trees were also observed in the median in the Boulevard Ring Road in Moscow. Many apple trees had been planted along both the Boulevard and Garden Ring Roads, but were removed when their apples began falling on passing automobiles. Apples trees may also be seen flanking Botanical Street along the Moscow Botanical Garden.

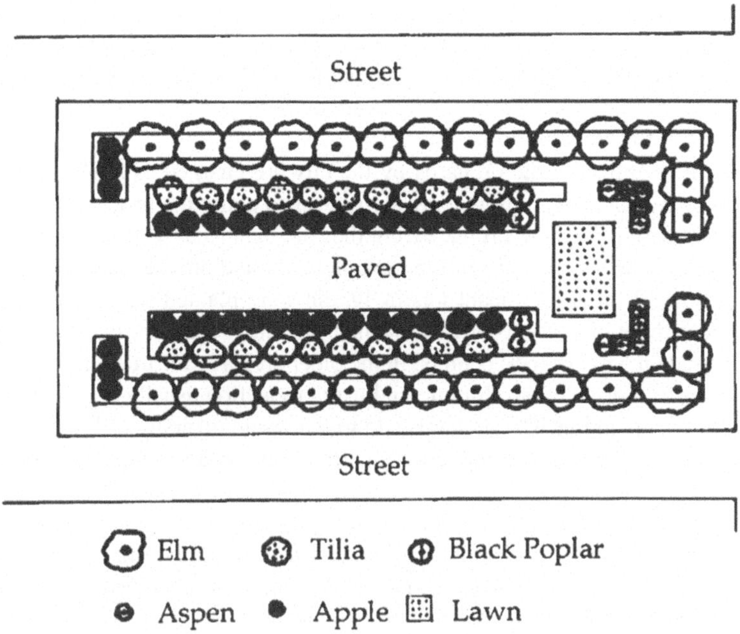

Fig. 7.4 Plan for small park with apple trees, Moscow

Several small parks in Moscow have been planted with apple trees either as individual specimen trees or in rows. A typical example was the trees planted in the small park adjacent to one of the Stalin Skyscrapers near the US Embassy (Fig. 7.4). Here apple trees were planted in raised beds along a central paved walk through the park. The trees were planted on 10 feet centers and have been pruned to encourage maximum fruit production.

Apple orchards have also been planted in the larger parks in Moscow. In Kolomenskoe Park an orchard covering one half acre was planted with rows 18 feet apart and with trees spaced 18 feet apart in the rows. Like apple trees observed in other urban landscapes in Moscow these tree were pruned to maintain a low crown canopy so that the apples could be picked from the ground using hooked poles. People frequently come here to gather apples when the fruit is ripe since the park is near a subway station.

Some apple orchards existing at the margins of Moscow were incorporated into an urban landscape of apartment buildings as the city expanded in the 1960s. In the Kolomenskoye area a section of an old apple orchard about 0.5 km long and averaging 50 m in width was retained. Five rows of trees in this orchard were preserved. These trees were planted on a 25 by 25 feet grid. The orchard provides apples for nearby apartment dwellers.

A unique aspect of the urban landscape of Tehran are the cherry orchards that have been planted by the Department of Urban Forestry on vacant lots owned by absentee owners. This practice, initiated during the early days of the Islamic

Republic of Tehran, is used to prevent illegal use of the vacant lots, reduce urban heat islands, control dust, reduce air pollution, and provide funds for the city (McBride and Mossadegh 2000). It is applied to larger vacant lots (greater than .25 ha; 0.6 acres). Many vacant lots in Tehran are owned by Iranians who left the country following the Revolution in 1979. The planting and initial maintenance costs incurred in establishing the orchards by the city are charged as a lien against the property to be paid if the absentee owner returns to claim the property. Once trees on these larger lots begin to produce fruit, they are managed by the city as commercial orchards to defray the cost of their maintenance and provide funds for the city.

Over 50% of the large vacant lots in Tehran were planted with cherry trees by 1999. Trees three to five cm (one to two inches) and two meters (six feet) tall had been planted on 5 m (16 feet) centers. In the older orchards the trees showed careful pruning for fruit production. Commercial fruit production usually began in the fifth year following planting. The orchards add to the visual diversity of local neighborhoods and contribute to the reduction of urban heat islands and air pollution in Tehran.

The production of apples and cherries from the urban orchards in Moscow and Tehran demonstrates the potential for urban orchards to supply fruit for urban dwellers and to contribute to the urban economy. Too often in other cities, especially in the United States, attempts to incorporate fruit and nut trees into urban forest designs on public property have been rejected on the basis of maintenance costs and problems of fruit litter. It is hoped that these examples of the use of fruit trees in Moscow and Tehran will encourage the establishment of more urban orchards.

Community Gardens

Community gardens were common to many of the cities surveyed in this study. Usually space in these gardens was devoted to raising of vegetables and flowers; however, larger urban gardens in some cities provided space for the incorporation of fruit and nut trees. Community gardens supporting trees were observed in Moscow, New York, Seattle, Stockholm, and Tokyo (Table 7.2). Cherry and apple trees were commonly observed in these community gardens with the exception of community gardens in Los Angeles. Winter temperatures in the Mediterranean woodland/scrub biome, where Los Angeles occurs, are not sufficiently low to induce abundant fruiting of cherry and apple trees. Fig trees were observed in community gardens in Los Angeles, a species not requiring low winter temperatures to induce flowering.

Narrow spaces along the smaller canals in Bangkok and adjacent areas of private yards are often planted with fruit trees to provide for local households. Trees noted along these canal including mango (*Mangifers indic*), jack fruit, papaya (*Carica papaya*), tamarind, kratin (*Leucaena leuocephala*), coconut (*Cocos nucifera*), mulberry (*Morus macrour*), and custard apple (*Annona squamosa*). These species were

Table 7.2 Tree species observed in community gardens

Biome	City	Tree species				
		Cherry	Pear	Apple	Plum	Fig
Broadleaf evergreen forest	Tokyo	•				
Deciduous forest	New York			•		
Coniferous forest	Seattle			•		
	Vancouver		•	•	•	
Mixed deciduous-coniferous forest	Moscow	•		•	•	
	Stockholm	•		•		
Mediterranean woodlandscrub	Los Angeles					•

• means the species is present

selected, planted, maintained, and their fruit harvested by individual homeowners, although they were frequently planted on public property along the small canals.

Storm Water Retention

Storm water retention basins provide temporary storage of runoff water to prevent flooding. They may also provide some treatment of pollution carried by storm water runoff. By holding storm water, retention basins increase infiltration of water and contribute to groundwater recharge. They are used to treat parking lot and rooftop runoff in urban areas. Many cities in the United States now require the construction of storm water retention basins as part of development projects with significant impervious surface areas.

Vegetation associated with storm water retention basins serve to absorb pollutants, provide wildlife habitat, and improve aesthetic quality. Trees play a role in all three of these functions. Although storm water retention basins depend primarily on grasses and aquatic herbaceous vegetation to remove pollutants, trees growing on the margins of these basins can also sequester and detoxify pollutants from storm water. Hybrid cottonwoods (*Populus trichocarpa* × *Populus deltiodes*), which are well adapted to the margins of storm water retention basins, have been demonstrated to oxidize trichloroethylene, a widespread industrial toxic element and to produce carbon dioxide and other metabolites (Strand et al. 1995; Newman et al. 1999). Willows (*Salix* ssp.) also have a demonstrated ability to cleanse contaminated sediments. They can reduce the concentrations of certain heavy metals (cadmium, copper, lead, and zinc) and organic contaminants (mineral oil and polyaromatic hydrocarbons) (Vervaeke et al. 2003).

Tree planting along with the planting of other vegetation can enhance the use of storm water retention basins by wildlife. In climates where storm water retention basins maintain minimum levels of water between rainfalls, these basins frequently serve as habitat for ducks and other types of water fowl. Shoreline plantings of wild millets (*Echinochloa* ssp.), bulrushes (*Scripus* ssp.), spike rushes (*Eleocharis* ssp.), and wild rice (*Zizania aquatica*) enhance the value of storm water retention basins

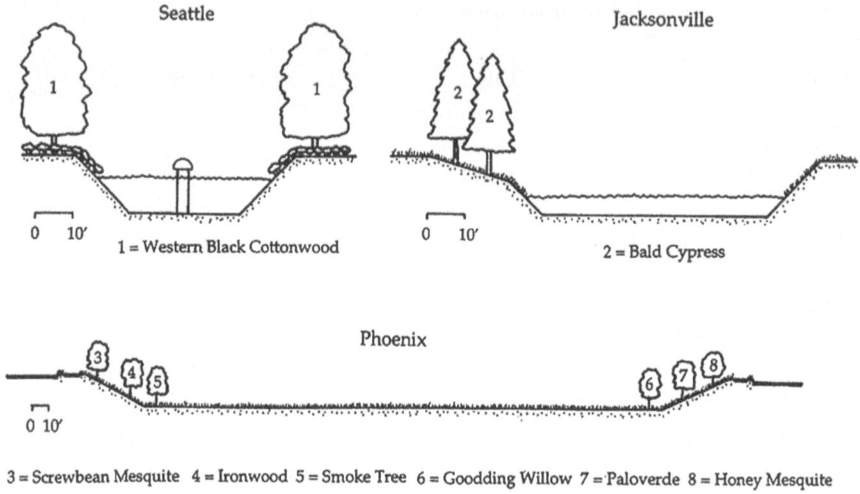

Fig. 7.5 Trees adjacent to storm water retention basins

for mallards (Martin et al. 1951). Mallards also supplement their diets by consuming acorns during the fall of the year. The incorporation of trees into the design of storm water retention basins not only enhances the diet of wildlife species, but also adds to the diversity of the habitat by providing niches for a variety of species.

Mozingo (1997) criticized the typical designs of storm water retention basin on the basis of their aesthetics. Referring to the use of plants in most projects as "greenwash", she called for more consideration of the potential aesthetic value storm water retention basins could have for local communities. Mozingo suggested that established principles of landscape design should be applied to integrate aesthetic values with pollution reduction and wildlife habitat values in the design of storm water retention basins.

Storm water retention basins were observed in Seattle, Jacksonville, and Phoenix. In all three cities trees were planted at retention basins and in all cases riparian woodland species native to the local biome were used. The contrasting pattern of tree planting in and adjacent to water retention basins in these three cities is illustrated in Fig. 7.5.

A storm water retention basin adjacent to the Meadow Brook Community Center in Seattle was excavated on the flood plain of Meadow Brook. It was designed for over bank flow from the creek as well as runoff from a nearby public school and community center. The excavated basin was irregular, forming a somewhat sinuous channel. The slopes down to the bottom of the basis were cut at an angel of 45° (1:1) and were planted with grass. At the tops of these slopes western black cottonwood (*Thuja plicata*), red alder (*Alnus rubra*), big leaf maple (*Acer macrophyllum*), Pacific black willow (*Salix lasiandra*), and Scouler willow (*Salix scouleriana*) trees had been planted. These trees were arranged in a naturalistic pattern to simulate riparian woodlands along streams in the Pacific Northwest. A variety of ground level herbs and shrubs of food value to wildlife were planted on the top of basin. The result was a natural looking setting. Mallards were on the water in the basin at the time the water retention basin was visited.

In Jacksonville a formal planting design was used in planting trees around water retention basins. Typical water retention basins in Jacksonville, like the one adjacent to Orange Park Mall, are rectangular in shape and have a 45° (1:1) slope from the floor of the basin to a height of six to eight feet and then a 26.5° (1:2) slope to a height of five to six feet above the break in slope. The slopes above the floor of the basin were planted with flood-tolerant species of grass. Bald cypress trees (*Taxodium distichum*) were planted in either one or two rows on the upper slope section with a between tree spacing of 45 feet. Bald cypress trees in these environments can reach heights of 100 feet and form a continuous crown canopy around the retention basin. During periods of maximum water discharge from the adjacent parking lots water in the basin will rise above the base of the bald cypress trees. This species is adapted to inundation of its root system and can survive temporary periods of flooding.

At the time of the observations made in Jacksonville all storm water retention basins contained water that was below the break in slope. Although the formal design did not attempt to simulate the appearance of the bald cypress swamps common to the Jacksonville area, it did depend on an understanding of the ecology of the species and its distribution in relation to gradients of inundation. It was a very appropriate design solution appropriate to the coniferous forest biome and the native forests in the Jacksonville area.

The water retention basins observed in Phoenix reflected the formal design seen in Jacksonville, but in less rigid planting patterns. Typical retention basins in Phoenix were excavated to a depth of 20 feet below the ground surface and had 1:2 slopes. Bermuda grass was planted on the floor of the basin as well as on the slopes. The floors of the water retention basins were being used as soccer fields when the retention basins were dry. In the desert climate of Phoenix the retention basins receive water during periods of intensive rainstorms. This water moves rapidly down into the sand and gravel substrate found beneath the floor of the basin. High rates of evaporation contribute to the loss of water. Trees were planted as individual specimen trees or in small groves on the slopes of the basin and occasionally at the edge of the basin floor. A toposequence simulated the arrangement one would find along dry washes in the nearby desert. At the lowest elevation, on or adjacent to the floor of the basin, desert willow (*Chilopsis linearis*), and smoke tree (*Psorothamnus spinosus*) were planted. On the middle sections of the slopes palo verde (*Cercidium floridum*) and ironwood (*Olneya testosa*) were used. Catclaw acacia (*Acacia greggii*) and mesquite (*Prosopis glandulosa*) had been planted on the upper slope to complete the toposequence. This arrangement of desert trees provide patches of shade at various elevations on the slopes of the retention basin and used trees in relation to their capacity to withstand different periods of inundation following desert storms.

Storm water retention basins will be required in areas of new urbanization in cities built on level topography that experience periods of intense rainfall. They may be established in areas of existing cities that have flood problems. These retention basins represent new urban environments where trees can be planted to add to the diversity and function of urban forests.

Storm water detention is achieved in many cities by the installation of bioswales. Strom water is diverted from city streets into swales that support herbaceous vegetation as well as trees. The plants in crease the surface roughness of the swales thus slowing down the water before it enters the storm sewer system. A portion of the water will percolate down into ground beneath the swale and never reach the storm sewer. The decrease in the velocity of the water results in the fall out of silt and particulate pollution. The plants in the swale can also contribute to the removal of pollutant from the storm water (National Association of City Transportation Officials 2013). Many tree species have been planted in bioswales throughout the United States and Great Britain. Species are selected on the basis of their capacity to withstand temporary inundation and their adaptation to winter temperatures in cities where they are planted (Scharenbroch et al. 2015). Examples of list of suggested species can be found in Jurries (2003) for Oregon cities, Bioswale (2012) for the northeastern United States, and Anonymous (2010) for San Francisco.

Brownfield Reclamation

The term "brownfield" has been applied to a site where soil and ground water has been polluted due to previous land use. Brownfields represent a special category of waste areas because of the contamination resulting from their former use. Former industrial sites and landfill areas constitute the most common brownfield sites. Federal programs in the United States and in other countries have focused on the reclamation of urban brownfields. Reclamation of contaminated soils on brownfield sites often requires off-site chemical-engineering solutions. Some contaminated soils can be reclaimed on-site through the use of plants. Phytoremediation can sometimes be achieved by planting trees on contaminated soils. The potential of various species of poplar (*Populus* spp.) and willow (*Salix* spp.) to remediate soils contaminated by heavy metals, trichloroethylene, mineral oil, and polyaromatic hydrocarbons has been reported in a number of field studies (Strand et al. 1995; Newman et al. 1999; Vervaeke et al. 2003). First year reduction in the concentrations of heavy metals were as high as 75 and 79% for organic contaminants. It should be recognized that some brownfield soils have contaminant concentrations that are prohibitive to the growth of trees. Trees recommended for phytoremediation are listed in Table 7.3.

Many urban brownfield sites are being reclaimed as sites for parks. Trees have been used in these reclamation projects both as park landscaping materials and for phytoremediation. In London, Camley Street Natural Park has been established on a brownfield site. This site previously housed a landfill and was contaminated by a variety of heavy metals and organic contaminants. Phytoremediation of the site has involved the use of crack willow (*Salix fragilis*), silver birch (*Betula pendula*), and

Table 7.3 Tree species commonly used for phytoremediation in the United States

Scientific name	Common name	Primary use
Liriodendron tulipifera	yellow poplar	Extracting organic compounds
Morus spp.	Mulberry	Extracting PAH (Polycyclic Aromatic Hydrocarbons)
Pinus taeda	Loblolly Pine	Extracting TCE (trichloroethylene)
Populus deltiodes	Cottonwood	Extracting organic compounds
Populus tremula	European aspen	Extracting organic compounds
Salix spp.	Willow	Extracting organic compounds
Tamarix spp.	Tamarix	Extracting PAH (Polycyclic Aromatic Hydrocarbons)

big tooth aspen (*Populus tremula*) as well as marsh species. The trees have been planted in small groves and as a margin to the marsh (Fig. 7.6).

As former contaminating industries and landfill operations are relocated outside of urban areas more brownfield sites will require reclamation. Trees play an important role in the biological remediation of these sites. Management of these elements in the urban forest require careful selection of tree species to maximize soil detoxification and the sequestering of heavy metals. The "life cycle" approach to the management of trees used for brownfield reclamation adds a new dimension to urban forest management.

Forest Products

Portions of urban forests in some cities function to produce timber and associated forest products in some cities. Seattle maintains watershed forest of over 90,000 acres that had been periodically harvested to produce income for the city (Williams 2004). This forest is dominated by Douglas-fir, western hemlock, western red cedar, and Sitka spruce and is a second growth forest that arose following extensive logging of the watershed in the nineteenth century. In 1986 41 million board feet of lumber were harvested from the forest and in 1994 receipts from timber sales amounted to $1,000,000 harvest for the city. In 1997 commercial logging was banned when the city adopted a long-term restoration plan for the watershed. Portions of the watershed continue to be logged as part of the restoration plan and proceeds from these harvests support a portion of the cost of the restoration.

Forest plantations occupy 5% of the area of the city of Tehran. First planted in 1935 to control dust, reduce air pollution, and arrest soil erosion, they now provide forest products and income to the city (McBride and Mossadegh 2000). The principal species used in these plantations are Eldarica pine (*Pinus eldarica*), Arizona cypress (*Cupressus arizonica*), and Persian ash (*Fraxinus oxycarpa*). Trees in these plantations have typically been planted on a 6 × 6 feet spacing (2 m × 2m) using bare root stock. Irrigation is provided to the trees by systems of contour irrigation

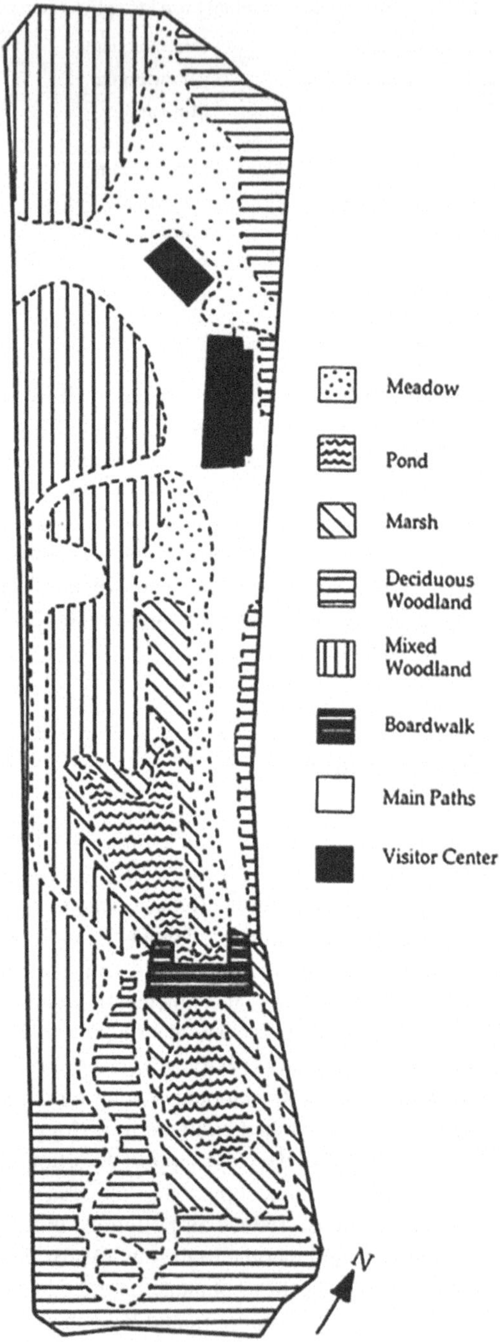

Fig. 7.6 Phytoremediation planting design for Camley Street Natural Park, London

ditches that are dug between every two rows of trees. The plantations are initially hand watered or watered by portable sprinkling systems during their establishment period (2–3 years). The principal forest products coming out of these plantations are poles that are harvested during thinning operations after the trees reach diameters of 4–6 inches (10–15 cm) at breast height. These poles are primarily used to support forms into which concrete is poured for floors and roofs during the construction of buildings. Hundreds of poles are needed to support the weight of concrete and reinforcing rods while the cement cures. The rapid growth of Tehran in the 1980s and 1890s created a large demand for construction poles. Pole production from the urban forest plantations of Tehran amounted to 56,000 poles in 1997 (Department of Greenspace 1998). It is anticipated that future products of these urban forest plantations will include saw timber and wood chips.

Urban forests stands in the larger parks in Moscow produce an abundance of mushrooms. These are harvested annually by the citizens of Moscow to be consumed in a variety of dishes. People in Moscow are allowed to pick the mushrooms without charge from forested parks in the city. No estimates are available as to the economic value of the annual mushroom harvest from the Moscow urban forest, but the value of these mushrooms to many Moscow citizens may be said to be 'bolshoi'.

Chapter 8
Management

Introduction

Management of urban forests involves tree establishment and maintenance activities necessary to carry out the designs and plans developed for different environments in an urban area. The design and planning responsibilities for urban forests rests in the same agencies that are tasked with tree establishment and maintenance in some cities, while these responsibilities are separated in other cities. Management of urban forests has been treated by Grey and Deneke (1978), Grey (1996), Miller (1997), Miller et al. (2015) and Roloff (2015) in their books on urban forestry. A review of their work will not be undertaken here. Instead, the purpose of this chapter will be to contrast administrative approaches to urban forest management observed in the cities studied, identify management problems confronting the urban forests of these cities, describe some special management activities observed, and identify alternative approaches in the management of the world's urban forests.

Administrative Approaches

Management of trees in urban areas falls to different municipal agencies in cities around the world. In many cities, such as Los Angeles and Paris, the responsibility for tree management is divided between the city parks department and the street department. This division of authority emerged in the nineteenth century with the creation of large urban parks. The difference in orientation of parks departments vs. street departments has led to different attitudes about urban trees and their management. A comparison of the lists of commonly used street trees with trees commonly used in small parks (Table 8.1) suggests different sets of selection factors are used in choosing trees. This is to be expected because of the different functions trees play in these two environments. However, the difference reflects what is often a

J.R. McBride, *The World's Urban Forests*, Future City 8,
DOI 10.1007/978-3-319-52108-4_8

Table 8.1 Common street and park trees observed in cities in the deciduous forest biome

Species			
Scientific name	Common name	Street tree	Park tree
Acer buergerianum	Trident maple		•
Acer campestre	Field maple	•	
Acer palmatum	Japanese maple		•
Acer platanoides	Norway maple	•	•
Acer saccharinum	Silver maple		•
Acer saccharum	Sugar maple		•
Aesculus hippocastanum	Horse chestnut	•	•
Ailanthus altissima	Tree of heaven		•
Betula pendula	Silver birch		•
Celtis occidentalis	Common hackberry		•
Cercis canadensis	Eastern redbud		•
Cercis siliquastrum	Judas tree		•
Crataegus monogyna	English hawthorn		•
Eucalyptus camaldulensis	Red gum		•
Fagus americana	American beech		•
Fagus sylvatica	European beech		•
Fraxinus excelsior	European ash	•	
Fraxinus pennsylvanica	Green ash		•
Gleditsia triacanthos	Honey locust	•	•
Ilex aquifolium	English holly		•
Juglans regia	English walnut		•
Liriodendron tulipifera	Tulip tree		•
Maclura pomifera	Osage orange		•
Magnolia × soulangeana	Saucer magnolia		•
Malus sylvestris	Wild crabapple		•
Paulownia tomentosa	Empress tree		•
Platanus × hispanica	London plane tree	•	•
Prunus avium	Wild cherry		•
Prunus cerasifera	Cherry plum		•
Prunus serrulata	Japanese flowering cherry		•
Quercus ilex	Holm oak		•
Quercus palustris	Pin oak		•
Quercus robur	English oak		•
Quercus rubra	Red oak	•	
Quercus velutina	Black oak		•
Robinia pseudoacacia	Black locust	•	•
Salix caprea	Pussy willow		•
Sorbus acuparius	Rowan	•	•
Sorbus intermedia	Swedish whitebeam		•
Syzygium panicultatum	Magenta cherry		•
Taxodium distichum	Bald cypress		•

(continued)

Table 8.1 (continued)

Species		Street tree	Park tree
Scientific name	Common name		
Taxus baccata	English yew		•
Tilia americana	American basswood	•	•
Tilia cordata	Littleleaf linden		•
Tilia × europaea	Common lime	•	•
Toona sinensis	Chinese toon		•
Ulmus americana	American elm		•
Ulmus procera	English elm	•	•

• means the species is present

difference in backgrounds of the individuals managing parks and those managing streets in urban areas. Park managers frequently come from backgrounds in recreation and social science while heads of street departments are likely to be drawn from engineering. This difference influences the way these individuals look at trees and their maintenance.

Some cities have combined responsibility for all the trees in a city under one agency. This appears to be the trend in the newer cities examined in this study. Murmansk, Reykjavik, Beijing, Tehran, and Dubai all have centralized the management of urban trees under one city agency. Beijing and Tehran are old cities, but each reorganized the management of urban trees under a single agency following their revolutions in the second half of the twentieth century. These cities named the agencies responsible for tree management with names that reflect a focus on plants (e.g., Beijing – Department of Landscape Gardening; Dubai-Department of Horticulture and Parks). The top administrators in these agencies were drawn from backgrounds in landscape architecture, horticulture, and forestry. How this difference in background will effect the development of the urban forests of these cities, as opposed to cities in which tree management agencies are headed by recreation experts and engineers is difficult to predict. The differences in management philosophy of the two groups may not be apparent for some time, since the urban forests observed today are legacies of decisions about urban design and tree management that stretch back more than 100 years.

A further dimension of the management of urban forests is who does the actual work on the ground. Many cities such as Tokyo rely on private contractors for the maintenance of cities trees and for tree planting. Periodic tree inventories, management plans for special environments, and designs for new parks are also contracted out in several cities. A few cities in the survey (e.g., Dubai, Tehran, Beijing) conduct all tree management activities with their own staff. Two thousand people are employed by the Department of Horticulture in Dubai to manage the municipal landscape and its trees. In Tehran a staff of slightly over 2000 gardeners manage street and park trees for the city. These workers are primarily drawn from the Horticulture Department of the University of Tehran. In Beijing, an army of 12,000 workers is employed in the management of street and park trees, as well as other urban greenery. About 9000 of these are laborers and 3000 are office workers.

The relative cost and benefits of managing urban forests with contract labor as opposed to city employees varies from city to city. In the United States there has been a trend to shift from city employees to contract workers as a cost-saving measure. The cost of benefits (health and disability insurance, retirement) required in the support of city workers has played an important role in decisions to move in the direction of outside contractors. The use of outside contractors has not been without unanticipated costs in some cities. Increased costs associated with city contracting procedures, scheduling of tree work, and quality control have presented problems for some cities.

Street tree maintenance is generally the responsibility of the city. San Francisco transferred the ownership and upkeep of trees to homeowners in 2100 after the recession cut into the city's budget. Many local homeowners could not afford the cost of tree maintenance and petitioned the city to allow them to remove trees. This was not allowed, which further angered some citizens. Over the next 5 years the quality of street tree care and sidewalk repairs due to root damage declined in San Francisco. Pressure was put on the Board of Supervisors to resume street tree maintenance. Measures to resume city responsibility for tree maintenance were debated by the Board of Supervisors. Much of the debate centered on the cost of the responsibility and how to fund the cost. A measure was put on the November, 2016 ballot to return the responsibility for tree maintenance, sidewalk damage, and liability to the city (Johnson 2016).

Urban Forest Management Problems

The municipal officials responsible for urban forests in the cities surveyed for this book were asked to identify problems they were facing in the management of their urban forests. Common responses were pest management problems and limited municipal budgets for urban forest management. In surveys conducted with urban forest managers in the United States by Beatty and Hickman (1981) and Kielbaso (1990), lack of adequate funding was identified as the largest single problem in tree management. Lack of sufficient water, nutrient deficiency, and vandalism were the top three problems associated with tree growth cited by urban forest managers in the survey conducted by Beatty and Hickman. Kielbaso identified mortality associated with tree planting, inadequate growing space, and under stocking as the major urban forest management issues, aside from inadequate budgets, in his survey. The fact that insect and disease problems were not highly ranked by urban forest managers in the United States in 1981 and 1990 compared to the high ranking in the cities surveyed for this book at the beginning of the twenty-first century is due to a change in the severity of urban tree pest problems.

Urban tree pest problems have increased in many cities during the last 10–20 years. Some of these changes have been brought about by exotic insects and diseases reaching cities on distant continents. In New York City an outbreak of the Asian long horned beetle (*Anoplophora glabripennis*) in 1997 cause damage to trees in parts of Brooklyn and Queens (Fig. 8.1) (Ghannam 1997). This epidemic resulted in the quarantining of parts of the city to prevent the transport of any potentially infected firewood or other woody material. The U.S. Forest Service supplied the city with $500,000 to help fund a tree removal program to control the insect. During the 1990s two exotic insects took a toll on eucalyptus trees in Los Angeles (Hagen 2004). The eucalyptus long horned borer (*Phoracantha semipunctata*) (Fig. 8.2) reached epidemic proportions in blue gum (*Eucalyptus globulus*) in the early 1990s to be followed by an outbreak of the red gum lerp psyllid (*Glycapsis brimbecombei*) in red gum (*Eucalyptus camaldulensis*) at the end of the decade. Urban forest pest problems have not been limited to American cities at the end of the twentieth century. In Tokyo concern was expressed over the pine wilt disease caused by an exotic nematode species (*Bursaphelenchus xylophilus*). The first outbreaks of this disease occurred in 1971 and the disease appeared to have peaked in 1998 (Thompson 2004; Yoshimura et al. 1999). The fusarium wilt of angsana (*Pterocarpus indicus*) caused a serious loss of angsana trees from 1994 to 1998 in Singapore (Fong and Lim 1999). In 2001 the horse chestnut leaf miner (*Cameraria ohridella*) invaded Paris as it rapidly spread over Western Europe (Guichard and Augustin 2002). This disfiguring leaf disease, coupled with the continued problems

Fig. 8.1 Asian long horned beetle

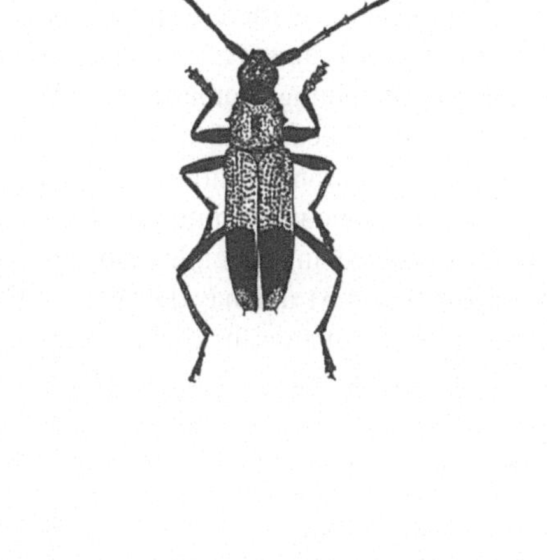

Fig. 8.2 Eucalyptus long horned borer

of leaf blotch (*Guignardia aesculi*) and anthracnose (*Glomerella cingulata*) in horse chestnut trees in Paris contributed to a high ranking of pest management problems in Paris in the survey. A pathogen, *Phytophthora ramorum*, was first described in Europe on ornamental species of *Rhododendron* and *Viburnum* (Harmon and Harmon 2004). In 2002 the pathogen was reported in California and Oregon. This pathogen causes the death of oak trees and infects other tree species as well as a number of shrubs commonly used in urban landscapes. Because of the rapid development of the disease in oak trees the malady is referred to as "sudden oak death". Kovacs et al. (2011) predicted the economic cost and property value losses attributed to the disease in urban areas in northwestern California at $7.5 million dollars for the treatment, removal, and replacement of more than 10 thousand oak trees. In addition they estimated property value losses to single family homes to be $135 million. International concern over the problems of invasive species at the close of the twentieth century also contributes to the high ranking given pest management issues by urban forest managers around the world (Liebhold et al. 1995).

Specific Management Activities

In preparation for the 2008 Olympic Games, Beijing followed an ambitious plan to add nine million trees to the city. Tree planting along with other landscape improvements was part of an effort to make the 2008 known as the "Green Olympic Games". To meet the objective of expanding the existing Beijing urban forest in a way that would make Beijing a "green city", a large percentage of the trees were planted along major boulevards and avenues were planted as large saplings. These saplings were six feet (2 m) long sections of the lower trunks of trees with a small portion of their root systems intact. The roots had been cut back to within six inches (150 cm) of the trunk after the trees were lifted from the tree nurseries (Fig. 8.3). The large saplings were cut from trees three to eight inches (8–20 cm) in diameter to facilitate ease of handling and transportation. The saplings were taken from trees in tree nurseries as well as forest plantations. No attempt was made to ball the meager root system or otherwise retain any soil in contact with the roots when the cuttings were transported for replanting. If the tree to be transplanted has branches below the 6 feet cut-off point, these branches were cut off at about nine inches (15 cm) from the trunk. In many cases there were no branches below the point at which the trees are cut. The large saplings were then planted in holes of sufficient size to accommodate their small remaining root systems. This system is an economical way to establish new street trees because of the low labor costs associated with the lifting of the trees in the nurseries, transportation, and planting.

This is not a new technique for the establishment of trees in urban areas. It was widely practiced in the eighteenth century in colonial towns in North America (Campanella 2003) and may date back to Greek and Roman times, used in Singapore for the planting of angsana in the twentieth century, and is currently also being used in Dubai as a means of saving ghaf trees. The uniqueness of the application of this

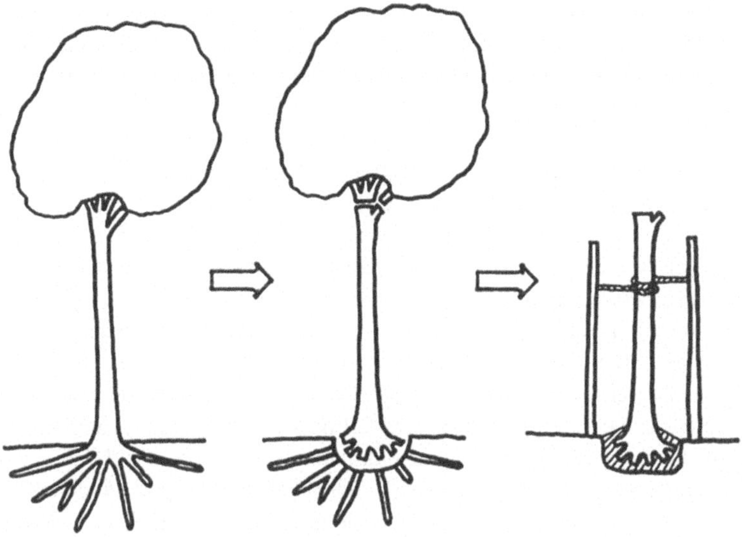

Fig. 8.3 Transplanting of large saplings

management technique for establishing trees was the scale at which it was used in Beijing. It is an economic technique that addresses the problem of shrinking budgets for urban tree management. It also serves as an inexpensive technique for the recycling of trees within urban areas where redevelopment projects eliminate existing tree cover. Some trees planted in Beijing, using the large sapling cutting technique, have been relocated from older hutongs that are being replaced by high rise apartment buildings. Not all of the trees in these hutongs are capable of being transplanted by this technique because of their size, but small trees are being relocated.

The large sapling technique does, however, present some problems for urban forest management. Among these are the rate of mortality, potential for limb breakage as the shoot system develops, and the reduce amenity value of streets planted in this fashion. An examination was conducted in 2002 of 750 large saplings recently planted in Beijing (Yang and McBride 2003). These large saplings of Japanese pagoda tree (*Sophora japonica)* and Chinese ash (*Fraxinus chinensis)* occurred along sections of Deiwai Street, Xueyuan Avenue, and Xueqing Avenue. The investigation found 25% of the transplants dead and another 31% were ranked as being in fair or poor condition. Chinese urban forestry officials pointed out that it is cost effective to replant the mortality under the labor market existing in Beijing and the cost of the large saplings. Time will be required to assess the potential branch brakeage problem. It is well known, however, that the strength of branch attachments from epicormic buds are weaker than that of branches arising from the apical mersitem (Harris et al. 1999). The amenity values of trees established by this technique was low in 2002 because of the stick-like appearance of the cuttings after transplanting. Even along streets where the cuttings have begun to branch the trees take on an unusual appearance. Pruning has improved the balance of the ill-formed crowns that developed in many of the large saplings. Re-evaluation of the large sapling technique used in Beijing after a decade should be undertaken to better judge its applicability for use in other cities. It may hold considerable promise as an economical management technique for the rapid establishment of urban forests.

The management of heritage trees represents a special urban forest management activity in some cities. Trees considered to be of heritage value because of their species, size, or historical significance are given protection under tree preservation ordinances in many cities. Eleven of the 33 cities in this study have heritage tree programs (Table 8.2). More of these programs occur in cities in the forested biomes. Heritage trees were initially recognized for their historic significance in the United States following the American Revolution (Campanella 2003). During the environmental movement of the 1970s the concept of heritage trees was expanded to include large trees, old trees, native species, trees with exceptional aesthetic value, and trees of ecological value. Many heritage trees fall into several of these categories, being at once of historic significance, large, old, native species of considerable aesthetic and ecological value. Heritage trees can contribute to civic pride and can be used in local history and environmental education programs. A variety of heritage trees were observed in this study.

Heritage trees of historical significance were located in Los Angeles, Stockholm, and Tokyo during the collection of data for this book. The Eagle Tree in Los Angeles was used by Spanish colonists to mark the northern boundary of a land grant known as Rancho San Pedro (Hodel 1988). The name "Eagle Tree" was given it by local resident for the great birds that formerly nested in the tree. It is a California sycamore tree of considerable stature. At 70 feet in height it towers above the surrounding residential area and can be seen from some distance. A small plot of ground around the tree has been fenced off to prevent soil compaction and to keep people away from the base of the tree. A bronze plaque identifies the tree and its historical significance. In Kungstrogarden (Opera Garden) in Stockholm a circle of eleven English elm trees are designated as heritage trees because of the efforts of environmentalists to save these trees from being cut down to build an underground station. People climbed into these trees in 1970 and remained for several weeks until the government reversed its plan to remove the trees. This victory is considered to be the start of the "Green Movement" in Europe. The historical importance of the eleven trees is noted by a sign and the trees are maintained as a prominent feature of the Opera Park. Several trees that survived the American fire bombing of Tokyo have been designated as heritage trees (Nagasaki 1998). Some of these have been designated by local wards within the city while other are on a national registry of historic trees. They are silent witnesses to the destruction of a large portion of Tokyo during World War II. After 60 years they exhibit large fire scars and distorted canopies as a result of the fires. The maintenance of these trees and the degree to which they are identified by signs varies with their locations. Especially large trees in conspicuous locations at shrines or on temple grounds are more frequently marked with signs than smaller trees in less conspicuous locations.

Trees of unusual size are designated as heritage trees in some cities. In New York City the largest specimens of about thirty species have been designated as Great

Table 8.2 Cities with heritage tree programs

Biome	City	Heritage tree program
Tropical forest	Bangkok	
	Rio de Janeiro	
	Singapore	•
Broadleaf evergreen forest	Hong Kong	•
	Sydney	•
	Tokyo	•
Deciduous forest	London	•
	New York	•
	Paris	
Coniferous forest	Jacksonville	•
	Seattle	•
	Vancouver	•
Mixed deciduous-coniferous forest	Beijing	•
	Moscow	
	Stockholm	
Mediterranean woodland/scrub	Athens	
	Los Angeles	•
	Tel Aviv	•
Savanna-thorn scrub	Brasilia	
	Lagos	
	New Delhi	
Grassland	Buenos Aires	
	Johannesburg	
	Tehran	
Desert	Cairo	
	Dubai	•
	Phoenix	•
Tundra	Murmansk	
	Nuuk	
	Reykjavik	
Highland	Cusco	
	Lhasa	•
	Quito	

• means the species is present

Trees by the Department of Parks and Recreation (Barnard 2002). These include the 63 inch DBH by 135 foot tall Hangman's Elm in Washington Square Park, a 66 inch DBH by 150 foot tall tulip tree in the Riverdale section of the Bronx, and an eastern cottonwood that has a diameter of 69 inches and a height of 120 feet in a residential neighborhood on Staten Island. These are trees of remarkable stature in comparison to trees typically encountered in New York City. Seattle is also blessed with heritage trees of grand stature. These include a Douglas-fir that is 27 feet in circumference (103 inch DBH) and 255 feet tall, a western red cedar with a circumference of

Fig. 8.4 Old ghaf tree incorporated into newly designed traffic circle in Dubai

20 feet (DBH = 76 inch) and a height of 150 feet, and western hemlock measuring 12 feet around (DBH = 46 inch) with a height of 150 feet (Jacobson 1989). All of these are located in city parks or the Seattle arboretum and are of natural origin. Not all heritage trees are necessarily giant trees. The threshold diameter for protected heritage trees in Jacksonville is only six inches for trees within 20 feet of a street right-of-way. In Sydney the threshold size is 300 mm (approximately 12 inches). In both of these cities, trees at or above these diameters are protected.

Old trees are revered in Beijing as in other parts of China. Aluminum signs, resembling automobile license plates are used to mark the old trees in the city. The plates are approximately two by five inches for trees over 100 years old, four by eight inches for trees over 500 years of age, and four by twelve inches for trees older than 1000 years. The signs call to attention the longevity of the trees and their important heritage value.

In Dubai, and throughout the United Arab Emirates, all ghaf *(Prosopis chineraria)* trees were given heritage status by the King Sheikh Zayed. Ghaf trees are common in the Arabian Desert in the Emirates and were historically a tree of many uses to local people. The designation of the ghaf tree as a heritage tree protects individual trees from being cut down. However, if urban development projects cannot accommodate existing trees in Dubai, those trees will be moved to other sites or to a special reserve for these trees, established in Mushref Park near the edge of the city. Here the trees are replanted and maintained indefinitely because of their heritage status. Several old ghaf trees within the city of Dubai have been incorporated into site designs and now serve as a component of the new urban landscape (Fig. 8.4). These trees are not designated by any signs, but are recognized by the people of Dubai as part of their heritage.

A further designation of special trees was observed in Bangkok where certain trees are wrapped with red or saffron colored "scarves" to signify the spiritual importance of these trees. Many of these trees became the repository of broken statues of Buddha that were left at the base of the trees along with flowers and incense. Broken statues of Buddha are not to be thrown away, but may be left at the base of these sacred trees. Great reverence is given to these trees by the people of Bangkok.

In many cities the management of heritage trees is limited to the protection given these trees by their designation as heritage trees. Designation of these trees identifies their importance to city planners and developers. Signage of these trees in cities like Beijing, Tokyo, London, and New York also contributes to their protection from ordinary citizens. In some cities more active maintenance programs are aimed at prolonging the life of the heritage trees. Techniques often used include the propping up of large lateral branches of trees in parks and temple grounds (Beijing, Tokyo), prevention of soil compaction around the base of tree by fencing (Los Angeles, Tokyo), loosening of soil around the base of heritage trees to increase oxygen access to roots (New York, Los Angeles), incorporation of fertilizers and organic matter into the soil around the base of trees (London, New York), and the transplanting of heritage trees faced with being cut down in areas of urban development (Dubai, Bangkok, Los Angeles).

Heritage tree programs have become battlegrounds for developers and environmentalists in several American cities. Real estate interests are sometimes stymied by tree preservation ordinances that prevent indiscriminate removal of trees from private property (Lloyd 2004). Some arborists and urban foresters have expressed concern over the use of tree diameters to designate heritage trees, because they fear, with some justification, that property owners may remove trees before the trees reach heritage size.

Alternative Approaches to Urban Forest Management

Five approaches to urban forest management were observed in the course of this study. They represent new ways of accomplishing the management of urban trees under the constraints of reduced municipal funding. These approaches are (1) corporate sponsorship of trees along boulevards and in city parks, (2) community tree planting programs, (3) non-government organization management of large urban parks, and (4) master arborist programs. Corporate sponsorship of trees along boulevards and in city parks was initiated in the 2001 in Singapore. In that year the Garden City Fund was started to provide opportunities for private individuals and companies to contribute financially and to participate actively in the maintenance and future development of the urban forest. It was intended to give Singaporeans a stronger sense of ownership of public parks and tree-lined boulevards. Since the initiation of the Fund a number of corporations and private citizens have stepped forward to fund specific projects such as the Heritage Tree Fund and the construction of children's garden at the Singapore Botanic Gardens. Corporations have adopted parks under the program. Park adoption involves a commitment of funding to support tree planting, design modifications, and maintenance. The employees of corporate sponsors and their families participate in annual events involving tree planting or other park improvement activities. These activities are coupled with environmental education programs that build a sense of pride in the sponsorship of the parks. Major highways and tree-lined boulevards are also under corporate sponsorship in Singapore.

Involvement of ordinary citizens in tree planting in urban areas dates back to the early nineteenth century in the United States (Campanella 2003). People were initially organized by leading citizens to help with the planting of trees in a number of small town and villages in New England. In 1872 J. Sterling Morten proposed a tree planting holiday in Nebraska to be called "Arbor Day" in an effort to revive interest in community tree planting and to contribute to the beautification of American cities (Everard 1926). The success of the first Arbor Day led to its adoption across the United States and in other countries. Arbor Days have resulted in the planting of millions of trees by members of local communities and youth organizations.

Community participation in the planting of trees reached a high point in Los Angeles on the twentieth anniversary of Earth Day when local people planted trees along seven miles of King Boulevard in honor of Martin Luther King, Jr. This community tree planting event was organized and supervised by a non-government agency known as TreePeople. A young man named Andy Lipkis first organized tree planting in 1973 in an effort to reforest areas in the mountains around the Los Angeles basin. He expanded the focus of his organization to include the planting of street trees in the 1980s and adopted the name TreePeople. TreePeople received a considerable boost in community support when it planted one million trees to improve air quality in Los Angeles in preparation for the 1984 Olympic Games. TreePeople as an organization grew to a staff of 40–50 full and part-time employees and a trained core of 500 volunteers by the beginning of the twenty-first century. They sponsor community tree planting programs in which they bring together neighbors for the selection and planting of street trees. Tree planting block parties are organized by TreePeople who provide trees, tools, expert instruction on tree planting, and quality control supervision. They enlist volunteer "citizen foresters" who are trained in tree pruning and maintenance to care for the newly planted trees. In addition to TreePeople's street tree planting programs, the organization launched a "fruit tree program" for low-income neighborhoods to supplement the diet of children. Approximately 3000 fruit trees (apples, plums, peaches, apricots, figs, and nectarines) are distributed annually to community groups concerned with food security in Los Angeles. After being trained by TreePeople in basic horticultural skills, volunteers from these community groups distribute and plant the trees with their clients, delivering a measure of future food security to group homes, community garden centers, shelters, orphanages, prison camps and housing projects.

The work of TreePeople in Los Angeles and similar non-government organizations in other cities have assisted in the management of trees in many cities in the United States and other parts of the world. This approach to management has been successful in augmenting limited municipal funds and staff for tree care and planting in urban areas. One concern raised with the relinquishing of municipal responsibility for street tree planting to non government organizations is the potential loss of control over some design aspects of the urban forest (Beatty 1999). Allowing individual homeowners to select the tree species they wish to plant in front of their house has usually resulted in considerable diversity in species composition. Although this may be viewed as democratic and ecologically sound, it can prevent the development of certain aesthetic and locational values street trees can provide.

The trade off between the effectiveness of non-government organizations in getting trees planted and overall municipal control of the design of the urban forest should be given further consideration.

The management of trees in a large urban park is exemplified by the management of Central Park in New York by the Central Park Conservancy. The Conservancy was founded as a not-for-profit organization in 1980 dedicated to the improvement of conditions in Central Park. Through a series of agreements and contracts with the city of New York, the Conservancy has assumed an increasing role in the management of the park. In 1998 an eight-year management contract was signed with the city that officially made the Conservancy "Keeper of the Park". The Conservancy has raised over $300,000,000 in support of a variety of restoration and management activities in Central Park. About one third of this amount has been set aside in an endowment to support future parks activities. Over $70 million, raised by the Conservancy, has been used to restore major park landscapes. These projects have included deferred tree maintenance and the planting of new trees, as well as work on lawn, shrub, and flowerbed restoration. Specific tree related activities have included an inventory of Central Park's more than 26,000 trees, development of a comprehensive management plan to manage the Park's three major woodlands, establishment of the Woodland Advisory Board to develop strategies and monitor the care of the park's 130 acres of woodlands, establishment of the Central Park Tree Trust to form an endowment fund to protect Central Park's trees, reintroduction of more than 40 species of native wildflowers, shrubs, and trees to the North Woods, restoration of the picturesque woodland and meadow-like garden at the Sixth Avenue and Central Park South entrance, establishing a monitoring program to detect the presence of the Asian long horned beetle in Central Park, planting of American elms along Fifth Avenue, and planting of more than 1400 understory trees and shrubs throughout the park. The Conservancy now provides more than 85% of Central Park's annual $20 million budget. Approximately four out of five Central Park employees are funded by the Conservancy.

The Conservancy has increased the horticultural staff of the park and enlisted thousands of volunteers to assist in special management activities. One of these activities is an annual removal of Norway maple seedling form the North Woods. A significant Norway maple invasion had taken place in the understory of this native woodland by the mid-1990s (McDonnell et al. 1997). A program aimed at controlling the invasion attracts over 100 volunteers each year to pull up the unwanted seedlings. As part of the woodland management effort in Central Park the Conservancy hired a woodlands manager and increased the woodland crew from two to five people in 1993.

The ability of the Central Park Conservancy to raise funds to augment city funds for the management of the Park serves as an example for other cities facing declining budgets for urban forest management. The Conservancy has also been able to attract leading experts to advisory boards, such as the Woodland Advisory Board, to seek the best advice on issues of tree management. The Conservancy has the capacity to respond to new issues facing the management of trees in Central Parks in an effective and efficient manner. The rapid response of the Conservancy in mustering volunteer support in monitoring park trees for the Asian long horned beetle is an example of this capacity. The Conservancy is a model of how a private organization can successfully manage a significant portion of an urban forest.

The Urban Greenery Society of Japan was established in 1947 to promote the management of trees and other vegetation in cities in Japan. Its membership includes urban foresters, landscape architects, arborists, and horticulturalists, as well lay people interested in urban green space. In 1991 the Urban Greenery Society initiated a program to protect and keep alive old trees in Japanese cities. The central concept of the program was to train and make available to cities in Japan technicians to work on problems associated with the management of old trees. The program has certified about 80 people a year as "tree doctors" in a 2-week long course. The training program has been staffed by the leading arborists in Japan who are specialists in the care of old trees. Staff of the Urban Greenery Society maintain a service for the translation of foreign publications on the care of older trees. These translations are supplied to their growing number of tree doctors. The Society has also created a "Tree Doctor's Association" to promote the program and communicate with tree doctors. The Association has annual meetings, special refresher training courses, and maintains a newsletter. The focus of this training is the improvement of soil conditions to promote healthy root systems in older trees. Japanese tree doctors trained by the Urban Greenery Society of Japan have volunteered their services to promote the longevity of trees on temple grounds, religious shrines, public parks, and on private property in Tokyo as well as in other cities in Japan. The capturing of expert knowledge from the best arborists in the country, translation of foreign publications on the care of older trees, and maintenance of an organization of trained volunteers are activities that could be applied in other cities to improve the management of older trees in urban forests around the world.

Chapter 9
Urban Forests, Biomes and History

The basic object of this investigation has been to understand the relationships between the characteristics of urban forests and the biomes in which they occur. Four major characteristics have been examined in relation to different urban environments common to all cities. These characteristics were urban forest composition, design, function, and management. Another aspect of the investigation has been to understand the influence of cultural and historical factors on the characteristics of urban forests. Some ideas about these relationships have been put forward in previous chapters on urban forest composition, design, function, and management. This chapter will be used to draw some general relationships between urban forest characteristics, biomes, and history

Species Composition

A total of 450 tree species were recorded in the 33 cities surveyed for this book. This number is far less than the total number of species that are being used in these cities. The sampling procedures did not explore in any detail larger cities parks where greater numbers of tree species are often found. For example, Parque do Flamengo in Rio de Janeiro support upward of 200 trees species while the sampling procedure used in this study recorded only 34 species. The number of species recorded in each city may, however, be used as an index to the relative tree diversity of the cities studied. In Vancouver the greatest number of species were recorded (64 species) and in Nuuk the least number (5 species) were observed. Based on these numbers and the numbers observed in the other cities, the species richness of Vancouver would be ranked number 1 and Nuuk 33.

A comparison of the ranking of native tree species richness in each biome with a ranking of the average species richness in the urban forest in these biomes is shown in Table 9.1. With the exception of the cities in the deciduous forest and tundra biomes, there is no correlation between the two rankings. Furthermore, when the

J.R. McBride, *The World's Urban Forests*, Future City 8,
DOI 10.1007/978-3-319-52108-4_9

Table 9.1 Ranking of tree species richness in biomes vs. urban forest

Biome	Biome tree species richness rank	Average number of tree species per city	Urban tree species richness rank
Tropical forest	1	42	5
Broadleaf evergreen forest	2	40	6
Deciduous forest[a]	3	48	4
Mixed deciduous-coniferous forest	4	35	8
Coniferous forest	5	50	3
Mediterranean woodland scrub[a]	6	56	1
Savanna/Scrub	7	54	2
Grassland	10	41	5
Desert	8	37	7
Tundra	11	10	10
Highland	9	35	9

[a]Deciduous forest given the rank of 2 because two of its cities had higher species richness that the two cities in the Mediterranean woodland/scrub with the highest species richness

biomes are grouped into three categories (tree dominated biomes, biomes in which trees are co-dominant with other plant life forms, and treeless biomes) and the average ranking of their native species richness is compared to the average ranking of the urban forest species richness of the cities within the biomes no trend emerges (Table 9.2). The highest ranking category in terms of urban tree species richness was the tree co-dominate biomes with a ranking of 1.5. The tree dominated biomes with an average native species richness ranking of 3.0 ranked in the middle position with regard to urban tree species richness. In contrast, those cities in the tree co-dominant biomes had the highest average urban tree species richness ranking. Cities in biomes that are generally treeless ranked lowest in terms of the species richness of their urban forests. It is apparent that the greater species diversity in the tree dominated biomes has not had a significant influence on species richness in cities in those biomes. In contrast the low tree species richness in the treeless biomes is reflected in the lower species richness of urban forests in these biomes. Climatic conditions that are unfavorable to tree growth in these treeless biomes is probably more significant in limiting urban forest species richness that the relatively low native tree species richness of the biomes. The high ranking of tree species richness in cities in the tree co-dominate biomes may be due to a relative dearth of tree species in these biomes. The absence of a large variety of local trees may have stimulated the desire for exotic species to enrich local urban forests.

Factors other than the diversity of trees species within a biome are important in the tree species diversity in the urban forest. This conclusion is borne out in when one looks at the average percentage of urban forest tree species in each of the tree dominated biomes that are not native to the biome trees (Table 9.3). Although the data indicates a relatively high percentage of trees in tropical forest biome cities are

Table 9.2 Comparison of rankings of biome tree species richness and urban forests tree species richness according to relative importance of trees in the biome

	Biome species richness	Urban species richness
	Rank	Rank
Tree dominated biomes:		
Tropical forest	1	5
Broadleaf evergreen forest	2	6
Deciduous forest	3	4
Mixed deciduous- coniferous forest	4	8
Coniferous forest	5	3
Average	3.0	5.2
Tree co-dominant biomes:		
Mediterranean woodland scrub	6	1
Savanna/scrub	7	2
Average	6.5	1.5
Biomes in which trees are of limited importance:		
Desert	8	7
Grassland	10	5
Tundra	11	10
Highlands	9	9
Average	9.5	7.8

Table 9.3 Tree species richness ranking of tree dominated biomes and average percentage of non-native urban tree species used in cities in these biomes

Biome	Tree species richness rank of biome	Average % of urban trees species not native to the biome
Tropical forest	1	37
Broadleaf evergreen forest	2	60
Deciduous forest	3	43
Mixed deciduous- coniferous forest	4	68
Coniferous forest	5	65

selected from that biomes, there is no correlation with the ranking of the species richness of the other biomes and the percentage of urban tree species used from other biomes.

The low urban forest tree species richness rankings of cities in the desert and tundra biomes is more logically related to climatic factors that to the species richness of the rankings of the two biomes. The low winter temperatures of the tundra biome limits tree species richness both in the biome and in cities within the biome. In attempting to identify tree and shrub species that could survive in the tundra climate, scientists at the Arctic Alpine Botanical Garden south of Murmansk have tested several hundred species. Similar test have been carried out by the Dubai Department of Parks and Horticulture to identify species that can endure the extremely high temperatures of Dubai. Irrigation can be used to supply water to trees and other landscape plants in Dubai, but adaptation to very high temperatures is the final determinant in what trees can survive. The relationship of urban tree species richness in cities within biomes noted for their extreme temperatures must not leads to any general conclusions about trends in overall urban forest species richness and biome climate. Although minimum temperatures play a predominant role in limiting the number of tree species that can be grown in cities in the tundra biome, the maximum temperatures encountered in cities in desert biomes has a less restrictive influence.

It is evident that trees common to both the native biome and the geographical region are used in local urban forests, although no strong correlation was discovered between the tree species richness of the biomes and urban forests. The percentage of trees native to the biome in which the city occurred varied from 0 to 79% (Table 9.4). The absence of trees in the grassland and tundra biomes accounts for the zero percentages recorded for cities in those biomes. Los Angeles recorded the highest percentage (79%) of trees native to its biome. The highest average percentage of native trees used in cities occurred in the Mediterranean woodland/scrub (56 percent) and deciduous forest biomes (48%). Examples of commonly used native species in the Mediterranean biome are Canary Island date palm (*Phoenix canariensis*), Italian cypress (*Cupressus sempervirens*), and olive (*Olea europaea*), while common lime (*Tilia europea*), green ash (*Fraxinus pennsylvanica*), English elm (*Ulmus campestris*), and pin oak (*Quercus palustris*) were frequently used in cities in the deciduous forest biome.

The reliance on native species in the Mediterranean woodland/scrub biome may be a function of the wet winters and dry summers typical of this biome. This rainfall pattern results in periods of water surplus and water deficit that is not experienced in any of the other biomes. The relatively high percentage of native species used in cities in the deciduous forest biome is related to the availability of numerous native tree species in this biome combined with low winter temperatures that limit the use of species from the tropical forest and broadleaf evergreen forest biomes.

The use of trees native to the region in which a city occurred showed a different pattern than that of species native to the biome. The geographical region in which a city occurs was defined by the adjacent geographical area. This area may include one or more biomes in addition to the biome in which the city is located. For example, the region defined in this analysis for Murmansk was the near-polar areas of Europe and Siberia. This geographical region includes portions of both the tundra and coniferous forest biome. The geographical region for each city and the percentage of trees observed in its urban forest that are native to the region are shown in Table 9.5. Examination of this table reveals that Murmansk supported the largest percentage of trees (92%) native to its geographical region in which it is located (the near-polar area of Europe and Siberia). Examples of tree species used in Murmansk that are native to its geographical region are Siberian larch (*Larix decidua ssp. siberia*), silver birch (*Betula pendula*), and rowan (*Sorbus aucuparia*). The use of trees in Murmansk that are native to its region is a result of the climatic limitations of the tundra biome and the experimental work that was conducted at the Polar-Arctic

Table 9.4 Percentage of trees native to the biomes of each city

Biome	City	Number of species	Species native to biome (%)
Tropical forest	Bangkok	36	75
	Rio de Janeiro	50	56
	Singapore	39	67
	Average	**42**	**66**
Broadleaf evergreen forest	Hong Kong	53	45
	Sydney	35	43
	Tokyo	33	33
	Average	**40**	**40**
Deciduous forest	London	46	56
	New York	59	70
	Paris	39	46
	Average	**48**	**57**
Coniferous forest	Jacksonville	41	27
	Seattle	45	33
	Vancouver	64	45
	Average	**50**	**35**
Mixed deciduous- coniferous forest	Beijing	35	26
	Moscow	40	25
	Stockholm	29	45
	Average	**35**	**32**
Mediterranean woodland/scrub	Athens	34	26
	Los Angeles	79	18
	Tel Aviv	48	25
	Average	**56**	**22**
Savanna/scrub	Brasilia	42	21
	New Delhi	54	17
	Lagos	25	16
	Average	**54**	**18**
Grassland	Buenos Aires	39	0
	Johannesburg	53	0
	Tehran	31	0
	Average	**41**	**0**
Desert	Cairo	38	21
	Dubai	28	25
	Phoenix	44	23
	Average	**37**	**24**
Tundra	Murmansk	13	0
	Reykjavik	12	0
	Average	**12**	**0**

(continued)

Table 9.4 (continued)

Biome	City	Number of species	Species native to biome (%)
Highland	Cusco	38	13
	Lhasa	18	11
	Quito	49	12
	Average	**35**	**12**
Overall average			34

Botanical Garden to identify trees that could survive and grow to arboreal form in the area. Over many years researchers at the garden transported trees from adjacent boreal forests of northern Europe and Siberia, as well as high elevation forests in Europe and North America, to test their ability to survive. This research led to the identification and adoption of many tree species in Murmansk. Other cities showing high percentages of the use of trees native to their geographical region are Stockholm (59%) and Tokyo (48%).

The average percentage of trees native to the geographical region in which the cities in each biome were located were low for the grassland biome (9%), desert biome (16%), and the Mediterranean woodland/scrub (16%). These low values are due to the limited number of tree species in the geographical areas in which the cities were located and the ease with which trees from other regions and biomes can be grown in these cities with the aid of irrigation. Cities in the mixed deciduous-coniferous forest and tundra biomes had the highest average percentages of trees native to the region, 51 and 50% respectively. These high percentages may be explained by the diversity of species available in the geographical regions surrounding cities in the mixed deciduous-coniferous biomes combined with relatively cold winters of the cities. In the case of the cities in the tundra the limits of the climate and the research done to identify appropriate species within the region accounts for the high percentage of trees used from the adjacent region.

Eleven tree species were frequently encountered as street and boulevard trees in several biomes (Table 9.6). This limited number of trees suggests in part the adaptability of certain species to urban environments across several biomes and the influence of historical factors on urban forest design. London plane tree (*Platanus acerifolia)*, the most frequently encountered species, is well adapted to urban environments and to the climatic variation among several of the world's biomes. This is due in part to its origin as a hybrid between the American sycamore (*Platanus occidentalis*) and the oriental plane tree (*Platanus orientalis*). The hybrid acquired a tolerance that allows it to survive throughout the range of its parent species. Further variation in the hybrid can be found in the dozen or so cultivars that have been developed since the original hybrid appeared in about 1645 (Jacobson 1996). London plane tree was first planted in an urban environment in London in the later half of the seventeenth century, but was not widely planted in cities until the early nineteenth century. Its presence as a frequently encountered street and boulevard tree in Johannesburg, Sydney, and New York City suggest the influence of British

Table 9.5 Percentage of trees native to the geographical region of each city

Biome	City	Region	Percent of trees used that are native to region
Tropical forest	Bangkok	Southeast Asia	25
	Rio de Janeiro	Brazil	26
	Singapore	Malayan Peninsula	28
	Average		**26**
Broadleaf evergreen forest	Hong Kong	South East China	21
	Sydney	Australia	38
	Tokyo	Japan and Adjacent parts of China and Korea	48
	Average		**36**
Deciduous forest	London	Western Europe	33
	New York	Eastern United States and Canada	42
	Paris	Western Europe	38
	Average		**27**
Coniferous forest	Jacksonville	Southeastern United States	46
	Seattle	Pacific Northwest region of the United States and Canada	18
	Vancouver	N. W. North America	20
	Average		**28**
Mixed deciduous-coniferous forest	Beijing	Northern China and Korea	57
	Moscow	Eastern Europe	38
	Stockholm	North Central Europe	59
	Average		**51**
Mediterranean woodland/scrub	Athens	Eastern Mediterranean	20
	Los Angeles	Southwestern area of California and adjacent parts of Mexico	13
	Tel Aviv	Eastern Mediterranean	21
	Average		**18**
Savanna-thorn scrub	Brasilia	Brazil	24
	Lagos	Africa	20
	New Delhi	India, Nepal, and Bhutan	42
	Average		**29**
Grassland	Buenos Aires	Central Argentina	10
	Johannesburg	Southern Africa	4
	Tehran	Northern Iran and adjacent parts of Russia, Afghanistan, and Iran	22
	Average		**9**

(continued)

Table 9.5 (continued)

Biome	City	Region	Percent of trees used that are native to region
Desert	Cairo	Northern Africa and adjacent parts of the Arabian Peninsula	8
	Dubai	Arabian Peninsula	14
	Phoenix	Southwestern United States and adjacent parts of Mexico	27
	Average		**16**
Tundra	Nuuk	Near-polar area of Eastern North America	0
	Murmansk	Near-polar area of Europe and Siberia	92
	Reykjavik	Iceland	17
	Average		**36**
Highland	Cusco	Andes Mountains	16
	Lhasa	Himalayan Mountains	11
	Quito	Andes Mountains	10
	Average		**12**

Table 9.6 The seven most frequently encountered street and boulevard trees

Common name	Scientific name	Number of cities[a]
London plane	*Platanus* × *hispanica*	7
Canary Island date palm	*Phoenix canariensis*	6
Norway maple	*Acer platanoides*	5
Jacaranda	*Jacaranda mimosifolia*	5
Rowan	*Sorbus aucuparia*	5
Black locust	*Robinia pseudoacacia*	4
Common lime	*Tilia europea*	4

[a]Where the species ranked among the top five street and boulevard trees on the basis of frequency of occurrence

colonialism on urban design and street tree selection during the early development of these cities. London plane tree was also introduced to Buenos Aires and Quito when they were European colonial cities.

Another example of the historical introduction influencing the composition of urban forests is the white mulberry tree (*Morus alba).* It was introduced in New York as well as many cities in the eastern United States to establish a silkworm industry in colonial America (Plotkin 2000). The silkworm industry failed, but white mulberry succeeded in being established as an urban tree. The spread of the silk making industry from Persia and China that began in the eleventh century resulted in a wave of planting of white mulberry across Europe (Lewington 1990). Although the species had been introduced earlier as a fruit tree and was found in a number of European medieval gardens that predate its introduction for silk production

(Landsberg 1995), the widespread planting of the tree for silk production contributed to its adoption as an urban tree. White mulberry was the most frequently encountered street and boulevard tree in Athens.

In the broadleaf evergreen forest biome both a dependence on (1) trees from the local flora and (2) historical and cultural factors play roles in the absence of commonality in street and boulevard species used in the cities studied. In both Sydney and Tokyo there is strong reliance on trees native to the region for street and boulevard planting. The broadleaf evergreen forests surrounding Sydney has a distinctly different flora from that of Tokyo. Sydney has employed native species as brown gum (*Eucalyptus robusta*), lemon bottlebrush (*Callistemon lanceolatus*), and rigid leaf (*Melaleuca styphelioides*) while Tokyo has utilized its native zelkova (*Zelkova serrata*), camphor (*Cinnamomum camphora*), and mochi tree (*Ilex integra*). The influence of England on Sydney is evident in the high frequency of London plane (*Platanus* x *hispanica*) along streets and boulevards while early influence of China on Japan can been seen in the use of gingko (*Ginkgo biloba*).

The lack of common species used for street and boulevard trees in cities in the savanna/scrub biome (Brasilia, Lagos, New Delhi) can be attributed to an interest on the part of urban planners to give a national identity to the cities through the use of tree species native to each country. Although New Delhi was designed by the two leading English architects of the day, Sir Edwin Lutyens and Sir Herbert Baker, their intent was to create a national capitol and relocate the seat of government from Calcutta (Tammita-Delgoda 1995). The use of tree species native to India was one way of creating a national identity for the new city. Brasilia was constructed in the late 1950s at a location far removed from other urban centers in Brazil. Its urban planning, architecture, and tree planting was revolutionary for its time. Selection of tree species for the initial planting of the city was influenced by the pioneering botanical work of Roberto Burle Marx. For many years prior to the design of Brasilia he had traveled through Brazil to identify native species that could be used in landscape design planting in Brazil (Montero 2001).

The occurrence of a relatively few prevalent species in various environments of the 28 cities studied suggest an analogy to the convergence of structural and functional characteristics observed in the evolution of plant species in geographically separated portions of a given biome. Mooney (1977) found similarities in the morphology, physiology, and ecological roles of plants occurring in the Mediterranean woodland/scrub biome in Chile and California. He considered these similarities to be the product of convergent evolution under similar climatic regimes in different parts of the world. Mooney believed that climate was the primary selective force driving convergent evolution. Humans have been the selective force (within the constraints of local climate) in a convergent selection of trees used in cities. When a species have been successful in an urban environment, such as any of the eleven prevalent species described above, it has been repeatedly planted in many cities within a biome and across biomes where climatic conditions permit. This dependence upon reliable species has sometimes contributed to a decrease in the diversity of species within the urban forests in a given biome. It may also account for the relatively high frequency of occurrence of a few species in most cities. Convergence selection of trees for planting in urban areas may also be illustrated in the selection of different species within the same genera in different parts of a given biome. An example would be the use of common lime *(Tilia europea)*, a European species, in cities in the deciduous forest biome in Europe and the use of basswood (*Tilia americana)*, a North American species common to the deciduous forest biome in North America. This convergence, through the selection of successful trees by urban forest planners, must account in part for the general absence of a strong correlation between urban forest composition and the characteristics of the biomes.

The primary conclusion to be drawn from this analysis with regard to urban forest species composition is that historical factors have been more significant in influencing composition than have been the characteristics of the biomes. The climate of the tundra biome does restrict the composition of urban forests within that biome. Climate may also limit the use of trees from the tropical forest biomes in other biomes subject to freezing temperatures, but no general trend in species composition was evident across the biomes. History, however, has made a more recognizable mark on urban forest composition.

Design

Design of urban forests considers the patterns used in tree planting. On a city wide basis design involves decisions about the planting of boulevards, streets, parks, cemeteries, and other urban environments. Comprehensive urban forest design can maximize the functional values of an urban forest by recognizing the importance of connectiveness, overall canopy cover, and the variations in urban topography (McPherson and Johnson 1988). At the site planning scale urban forest design addresses the placement of individual trees, their arrangement, and spacing. Design parameters reviewed by Arnold (1993) can lead to beautiful as well as functional designs for trees along streets, in parks and urban plazas.

Landscape architects and urban designers frequently look at the regional setting of a site for context to inform their design. This context may be ecological, historical, and/or social. The designs observed in the world's urban forests often reflect a context stemming from the characteristics of the biomes as well as historical and social factors. The contextual aspects of urban forest design based on the characteristics of the biomes is most evident in those biomes characterized by extremes in climate. The patterns of tree planting characterized by close spacing and multiple trees per planting spot observed along boulevards in Murmansk were designed to reduce sidewalk wind velocity and minimize wind chill in the tundra biome. Pruning practices resulted in almost hedge-like rows of rowan and birch along many thoroughfares in the city. When asked why multiple trees were planted in each planting spot local urban foresters replied that it was done to simulate the common pattern of clumps of trees observed in the boreal forest south of the city. These natural clumps are the result of the browsing of young seedlings by forest herbivores.

Tree spacing along streets did not correspond to any evident characteristics of the biomes. However, increased spacing was observed between trees in the medians of boulevards when the biomes were grouped into three categories based on tree dominance. The average of the modal spacing increased from 35 feet in tree dominated biomes to 40 feet in biomes in which trees shared dominance with other plant life forms, to 42 feet in treeless biomes. This increase is related in part to irrigation requirements in the biomes that are not tree dominated. There also appears to be a significant historical influence on the spacing of trees in boulevard medians when one compares cities whose urban forests were initially developed under the influence of colonial powers. For example, the modal boulevard tree spacing in London is 40 feet. This modal spacing was also observed in New York, New Delhi, Cairo, Singapore, Sydney, and Vancouver. In Beijing modal boulevard tree spacing is 20 feet. This spacing was used in Lhasa in the design of new boulevards following the take over of Tibet by China.

Shading of streets and sidewalks is a function of the design and can be influenced by the maintenance of street trees. Close spacing of trees will result in a continuous canopy over a sideway and the selection of species with long horizontal branches could provide for shading of street surfaces. However, pruning practices can prevent tree canopies from closing along and across a street. It might be assumed that the level of solar radiation associated with each biome could be a predictor of the degree of canopy closure. Those biomes that experience high levels of solar radiation could be assumed to be biomes in which tree canopy cover along and across streets would be maximized. The desert biome receives the highest level of solar radiation but ranked lowest in tree canopy cover along and over street and boulevards (Table 9.7). The low percentage of street and boulevards in cities in the desert biome that had canopy cover over their sidewalks and streets is due to the necessity for irrigation in the desert biome. Shade is very much appreciated in cities in the desert, but irrigation costs and the marginal reduction in air temperature resulting from shading appear to have not justified street tree planting designs that shade streets and sidewalks. Cities in the grassland biome exhibited the greatest percentage of streets and boulevards where the planting design resulted in the shading of sidewalks and streets. The urban forests of these cities provided an impressive contrast to the grassland that surrounded the cities. Shade is valuable in these cities but the ranking of the relative level of solar radiation received in these cities would not suggest the highest ranking for shading of streets and sidewalks. The second overall highest ranking for shading of streets and sidewalks occurred in cities in the tropical forest biome and the mixed deciduous –coniferous forest biome. The greater shading in cities in the tropical forest biome does respond to the high radiation levels experienced in these cities, but the annual radiation levels received in cities in the mixed deciduous-coniferous forest biome do not correlate with the percentage of streets and sidewalks shaded by trees. The data on canopy cover along and across streets, like the data on tree spacing correlates very weakly with the characteristics of the biomes.

Tree canopy cover in small parks is also not correlated directly with solar radiation. Small parks in cities in the desert biome showed the lowest percentage of tree

Table 9.7 Tree canopy cover over sidewalks in relation to annual solar radiation

Biome	City	Annual solar radiation (kg-cal/cm^2/year)	Tree canopy cover over sidewalks (%)
Tropical forest	Bangkok	150	76
	Rio de Janeiro	135	84
	Singapore	150	56
	Average	**145**	**72**
Broadleaf evergreen forest	Hong Kong	135	74
	Sydney	140	69
	Tokyo	110	54
	Average	**128**	**66**
Deciduous forest	London	90	47
	New York	120	50
	Paris	110	88
	Average	**110**	**62**
Coniferous forest	Jacksonville	145	44
	Seattle	110	50
	Vancouver	110	65
	Average	**122**	**53**
Mixed deciduous-coniferous forest	Beijing	125	80
	Moscow	90	92
	Stockholm	90	12
	Average	**102**	**61**
Mediterranean woodland/Scrub	Athens	160	50
	Los Angeles	160	36
	Tel Aviv	180	41
	Average	**167**	**42**
Savanna/scrub	Brasilia	150	12
	Lagos	170	50
	New Delhi	170	89
	Average	**163**	**50**
Grassland	Buenos Aires	140	94
	Johannesburg	160	66
	Tehran	150	83
	Average	**150**	**81**
Desert	Cairo	200	32
	Dubai	190	40
	Phoenix	180	16
	Average	**190**	**29**
Tundra	Murmansk	60	50
	Nuuk	70	0
	Reykjavik	70	0
	Average	67	17

(continued)

Table 9.7 (continued)

Biome	City	Annual solar radiation (kg-cal/cm^2/year)	Tree canopy cover over sidewalks (%)
Highland	Cusco	175	23
	Lhasa	180	57
	Quito	170	4
	Average	**175**	**28**

canopy cover. Canopy cover in these parks, particularly in the Middle East, is kept low to allow for rapid re-radiation of heat at night. The absence of a tree canopy cover allows heat absorbed during the day to rapidly dissipate. The "greenhouse" effect that keeps the air beneath a forest canopy warm at night is eliminated from small parks with little or no tree canopy cover. As a result these parks become cool oasis in the evening and are used at that time in many desert cities in contrast to the minimal use they receive during the heat of the day.

The 16 patterns observed in the placement of trees in small parks showed no correlation with the characteristics of the biomes. These designs were observed across a wide range of biomes and could not be directly tied to any determinants dictated by the biomes. The possible exception being the absence of grid and grove patterns in small parks in desert biomes. These patterns would interfere with the re-radiation of heat at night and were therefore avoided.

The tree planting designs observed in large urban parks, housing areas, urban greenbelts, water's edge environments, cemeteries, freeways, and roof top gardens also show little influence from the characteristics of the biomes. Historical, cultural, and social factors have influenced the way in which trees are employed in the designs for these urban environments. Temperature and water deficit may limit the pallet of plants that can be used in some biomes in the design of these spaces, but the patterns used in designing these spaces are seldom constrained by the characteristics of the biomes.

Function

The functions of the urban forest are not limited to biomes. The characteristics of the biomes may limit the tree species that can be used in a specific biome to fulfill a functional role. However, each biome with the exception of the tree-less biomes has a variety of tree species that can be used in specific functional roles. Air pollution reduction by an urban forest is maximized by trees with large leaf surface per unit of crown volume. Trees species with large leaf surface area per unit of crown volume are available within each of the biomes that support trees. All of the cities surveyed supported trees from other biomes. Many of these commonly used species, such as London plane tree, green ash, and European linden are very effective at reducing air pollution levels. Climatic conditions in the tundra and desert biomes will limit the effectiveness of trees to reduce air pollution levels. The short growing season and the long period of time deciduous trees are without their leaves in cities in the tundra biome will limit to amount of air pollutants than can be removed by trees on an annual basis. In the desert biome water deficit minimizes transpiration. Many species transpire for only a limited period of time in the early morning. Transpiration is important to the process of the removal of air pollutants by trees. Gaseous pollutants are removed from the air by both adsorption onto the surface of leaves and absorption into the leaves via the stomata. The water deficit and high evaporative stress in the desert biome minimizes a trees capacity to absorb air pollutants by reducing both the period of time stomata are open and the aperture of the stomatal pore.

Landslide stabilization observed in Monterey pine and eucalyptus plantations around Quito is being accomplished in a number of other cities through the planting of these and other species. In watersheds around Los Angeles area a variety of native and exotic pine species have been used. In the mountainous outskirts of Tokyo plantations of Japanese red cedar (*Cryptomeria japonica*) has been planted, while in mountains surrounding Beijing a variety of Chinese forest species have been used for erosion control and landslide protection. As with the use of trees for air pollution reduction, each biome can support a variety of tree species that can be used for landslide protection.

Urban orchards can and have been be established in most of the biomes. The climates of particular biomes will limit the species that can be used to produces fruit and nut crops. Tropical species such as bananas, jack fruit, and mangoes cannot survive freezing temperature found outside of the tropical forest biome, while apples and cherries will not get enough low temperature exposure to produce fruit in many of the warmer biomes. Regardless of these biome imposed limits there are trees species adapted to each of the biomes, with the possible exception of the tundra biome, that can be used to establish urban orchards.

The use of trees to enhance the appearance of storm water retention basins and to serve in phytoremediation projects is also not limited by the characteristics of the biomes with the exception of the limit a biome's climate may place on the use of any particular species. Willow species abound in nearly all of the biomes and serve very well in these functional roles. Other riparian species native to the geographical region of a city can also serve in these newer functional roles. Several native riparian species have been used in bioswales in the Pacific Northwest region of North America (Jurries 2003; Greater Vancouver Sewerage & Drainage District 2012).

The production of forest products from urban forests has occurred across several biomes and is not a biome limited function. Like the functional roles of the urban forest discussed above, the influence of the biome may be to some degree at the level of species selection for production forests in urban areas. Many species of eucalyptus and pine can be grown in several biomes and have become the species of preference for the production of forest products in many urban forests.

It must be concluded that function is not a biome limited aspect of the urban forests. Each biome supports native species that can serve in a variety of functional roles. Urban forest functions can also be achieved by using tree species introduced from other biomes.

Management

Some characteristics of certain biomes have rather direct implication for the management of urban forests. Water deficits encountered in cities in the desert biome and the extended period of summer drought common to the Mediterranean woodland/scrub biomes require irrigation for the establishment and maintenance of many tree species. Extremely low temperatures in the tundra biome require the mulching and protective covering of trees seedling during their first winter and the occurrence of permafrost may require special planting techniques.

Biome characteristics may also play a role in the increasing pest problems being encountered in the urban forests of the world. Exotic tree pests, such as the eucalyptus long horned borer and the emerald ash borer, have succeeded in part in urban forests due the absence of the predators and parasites that control their population in their native habitat. When tree species like blue gum (*Eucalyptus globulus*) are introduced into urban forest in cities in different biomes from the native biome of the species, it can be expected that tree pests that may subsequently arrive from the native habitat will not encounter predators and parasites that normally would keep them in check. This was the case with the eucalyptus long horned wood borer that destroyed large numbers of blue gum trees in southern California until parasitic wasps were introduced from native habitats in the broadleaf evergreen forest biome in Australia (Paine et al. 1995). This problem may be exacerbated if the climatic differences between the biomes add to the stress conditions of the tree species. Blue gum trees planted in the Mediterranean woodland/scrub biome of southern California experienced drought stress during the dry summers that made these trees particularly vulnerable to attack by the eucalyptus long horned borer.

The administrative approaches to urban forest management, special management activities, and alternative approaches to the management of urban forests discussed in Chap. 8 are independent of biome constraints that influence tree species selection, irrigation requirements, and winter protection of recently planted trees. Historical, cultural, and social factors may influence the administrative approaches to the management of urban forests. Citizen involvement in many urban tree management activities, such as tree planting, is common in many, but not all of the cities studied. Likewise, corporate and non-government organization sponsorship and management of urban parks has not universally been adopted in many of the world's urban forests due to historical presidents for the management of urban forests.

Conclusions

The general conclusion that may be drawn from this analysis of urban forest composition, design, function and management in relation to biomes and history is that history has played a far more important role in shaping the characteristics of urban forests. The characteristics of the biomes, especially climatic limitations to tree growth encountered in the desert and tundra biomes, have been important constraints on urban forest composition and some management practices in cities in these biomes. But, climate has been less limiting to the selection of tree and management practices in most of the other biomes. No general correlations between species richness of urban forests and biomes could be established. Similarly, the design characteristics of urban forests do not reflect biome characteristics except in those biomes at the margins of tree growth. Historical precedents for the composition, design, function, and management of urban forests have been more important than biome characteristics in shaping the urban forests of the world.

Chapter 10
The Future of the Urban Forest

The future of the urban forest is a relevant issue to ponder. Much speculation was generated at the close of the twentieth century concerning how cities and urban life would evolve over the next 100 years. Changes in world economics, communication, international politics, migration, and lifestyles in the 1990s suggest that the twenty-first century will be as different from the twentieth century as the twentieth century was from the nineteenth century. How these changes will effect the world's urban forests are difficult to predict, but an examination of recent trends in the characteristics of the urban forest, management of urban trees, current research, and the evolving nature of cities allows for some fertile speculation.

Space in urban areas for trees and civic devotion to tree planting may have reached a zenith in the later half of the twentieth century. Urban development in the United States following the Second World War was characterized by tree-canopied suburbs, new tree-studded industrial campuses, and the insertion of trees into formerly treeless neighborhoods. To some degree this pattern of increased tree density and canopy cover also occurred in Europe and in parts of South America, Africa, and Asia. However, at the beginning of the twenty-first century a decline in tree canopy cover in American cities was reported by the Urban Forest Center at American Forests (Brittin 2003). This report suggested a decline of 30% in urban tree canopy cover in the last three decades of the twentieth century. The decline is in part due to the removal of larger trees in urban areas and their replacement by young, smaller trees. A more alarming contribution to the decline is the failure to replace trees that are cut down. Estimates of only one tree replanted for every four trees cut down in American cities do not bode well for the future of the urban forest (Lipkis and Lipkis 1990). Even more drastic figures are reported for African cities where for every 29 trees being cut in urban areas only one is being replanted (Trees for Life 2004).

There are examples of accelerated tree planting programs taking place at the beginning of the twenty-first century that will increase tree numbers and canopy cover in a significant way for some cities. Both Athens and Beijing were actively involved in tree planting programs in preparation for the Olympic games that took

J.R. McBride, *The World's Urban Forests*, Future City 8,
DOI 10.1007/978-3-319-52108-4_10

place in the first decade of twenty-first century. Athens set a goal of planting a "Global Olive Wreath" to encircle the world along the Olympic Torch route for the 2004 Olympics. The trees were planted in 50 of the world's cities. Beijing planted nine million trees in preparation for the Olympic there in 2008. The Olympic games have been a rallying point for tree planting in host cities in the past. Los Angeles planted one million trees in preparation for the 1984 games while two million trees were planted for the 2000 Olympic games in Sydney in 2000. Tree planting for the Olympic Games has a long history dating back to the planting of an olive tree by King Iphitos of Elis. Tradition suggests that he was instructed to plant the tree by the Delphic Oracle. The leaves of the tree were to supply wreaths for the victorious athletes (Ridington 1935). One may anticipate that future Olympics will stimulate tree planting in host cities, but these localized efforts may not curtail the overall decline in tree numbers and canopy cover in the world's urban forest.

Tree planting campaigns, such as Global ReLeaf initiated in 1989, set the ambitious goal of planting 20 million trees for the new millennium. Not all of these trees were to be planted in urban areas, but programs like Global ReLeaf may help cities with declining budgets for urban forest management to move forward in addressing the imbalance between the number of trees removed and the number of trees replanted. The Global ReLeaf program brings together non-profit organizations, city governments and corporate sponsors to stimulate tree planting. The long-term effectiveness of this and other tree planting programs that bridge non-profit organizations and city governments depends heavily on the quality of leadership of the organization and the ability of the programs to accomplish their stated objectives. The continued success of programs like Global ReLeaf and the Los Angeles based TreePeople give one hope that the current decline in tree canopy cover and tree numbers can be reversed during the early decades of the twenty-first century.

Any attempt to project urban tree canopy cover and tree numbers in the twenty-first century must recognize the potential natural and political disasters have for tree populations. The twentieth century saw an extensive loss of elm trees from the urban forests in the deciduous forest biome in Europe and the United States. This disease was caused by the fungal agent *Ophiostoma ulmi* and is believed to be of Asian origin (Holmes 1978). In spite of well-funded federal research and management programs the disease practically eliminated the American elm tree from the urban forests of the eastern United States (Campanella 2003). Its impact was especially significant for a large number of small towns and villages in New England that depended almost exclusively on the American elm as a street tree. Currently there is considerable concern over the potential of the emerald ash borer (*Agrilus planipennis*) to destroy populations of various ash species in urban forests in the North America and Europe (USDA 2002). How the emerald ash borer and other exotic species of insects and disease will shape the urban forests of the twenty-first century is difficult to predict.

Hurricanes, typhoon, and earthquakes devastated urban tree populations in the twentieth century and in the early part of twenty-first century. In 1938 a hurricane struck the coast of New England causing extensive wind throw and wind breakage to trees in cities and forests from Long Island to southern New Hampshire and

Vermont. The city of New Haven lost 13,500 trees, Harford 100,000, and Boston 8000 (Felt 1939). Tokyo experienced typhoons and earthquakes that destroyed large number of street and park trees in the twentieth century (Cheng et al. 2000). The first major disaster to impact the Tokyo urban forest in the last century was a typhoon in 1917. It resulted in a loss of more than one-half of the city's trees. A second major disaster, the Kanton earthquake in 1923, caused the loss of 14,300 (60%) of the street trees. An additional 10% of the street trees were cut illegally after the earthquake due to the shortage of fuel. Efforts to replant street trees following the 1923 earthquake were completed by 1933 when the Tokyo street tree population again reached 33,700. In 1938 another typhoon destroyed about 39% of the street trees in the city. Strong winds blew down 18,200 trees while 12,500 trees were partially blown down, and an additional 1700 trees had broken trunks. A large percentage of the surviving trees had been tilted by the high winds (Greenery Research Laboratory 1981).

In August 2005, Hurricane Katrina struck the coastline of Louisiana, Mississippi, and Alabama with winds in excess of 150 miles per hour. This wind, coupled with a tidal surge, toppled thousands of trees in the path of the storm in dozens of cities and towns as it moved inland. Further tree mortality occurred in New Orleans as a result of prolonged flooding of 80% of the city. Hurricane Katrina is ranked as one of the most costly natural disaster to strike the United States. Tree damage in cities caused by ten hurricanes coming ashore in Florida, Louisiana and Puerto Rico from 1992 to 2005 was studied by Duryea et al. (2007a, b). They ranked 192 tree species on the basis of their capacity to withstand hurricane force wind. Duryea and her associates found an expected relationship between wind speed and percentage of the urban forest loss. In hurricane Erin (1995, 85 mph wind velocity) a little over 10% of the urban forest was lost in towns along the Florida panhandle. Hurricane Andrew, with winds of 165 mph, was responsible for the destruction of 35–40% of the urban forests in cities in southeastern Florida. Tree susceptability to hurricane force winds was related to wood strength, flexibility, pattern of root growth, tree age, and branching pattern. Species with the highest wind resistance include oak (*Quercus* spp.), hickory (*Carya* spp.), holly (*Ilex* spp.), podocarpus (*Podocarpus* spp.), cypress (*Taxodium* spp.), Phoenix palms (*Phoenix* spp.), pindo palm (*Butia capitata*) and sabal palm (*Sabal palmetto*).

Global climate change has the potential to modify the composition, design, structure, and management of the world's urban forests in the twenty-first century. Various global climate change models suggest local patterns of increasing temperature will be accompanied by either increasing or decreasing precipitation (Schneider 1993). Predicted changes in precipitation for the cities surveyed for this book are shown in Table 10.1. Twenty-one of these cities are expected to have increased annual precipitation while the remaining six will experience decreased precipitation. The consequences of these changes for urban forest species composition will depend upon the magnitude of the change. Davis and Zabinski (1992) predicted the elimination of sugar maple (*Acer saccharun*) and eastern hemlock (*Tsuga canadensis*) from the area around New York City using the United Kingdom Meteorological Office (UKMO) model. Other species now common to the New York urban forest

Table 10.1 Predicted changes in temperature and precipitation due to global warming for cities studied

Biome	City	Temperature increase (annualC)	Precipitation increase (mm/day)
Tropical forest	Bangkok	3–5	5–7.5
	Rio de Janeiro	3–5	5–7.5
	Singapore	3–5	5–7.5
	Average	3–5	5–7.5
Broadleaf evergreen forest	Sydney	3–5	5–7.5
	Tokyo	3–5	5–7.5
	Hong Kong	2–4	−0.1 to −03
	Average	2.7–4.7	5–7.5
Deciduous forest	London	2–3	1–3
	New York	3–5	1–3
	Paris	3–5	1–3
	Average	2.7–4.3	1–3
Coniferous forest	Jacksonville	3–5	3–4
	Seattle	3–5	3–4
	Vancouver	2–4	3–4
	Average	2.7–4.7	3–4
Mixed deciduous-coniferous Forest	Moscow	3–5	1–3
	Beijing	3–5	1–3
	Stockholm	3–5	1–3
	Average	3–5	1–3
Mediterranean woodland/scrub	Athens	3–5	1–3
	Los Angeles	3–5	0–1
	Tel Aviv	4–5	−0.5 to −0.1
	Average	3.3–5	0.25–1.3
Savanna/thorn forest	Brasilia	5–10	5–7.5
	Lagos	3–5	0–0.5
	New Delhi	3–5	1–3
	Average	3.7–6.7	2–3.7
Grassland	Buenos Aires	2–3	1–3
	Johannesburg	3–5	1–3
	Tehran	3–5	0–1
	Average	2.7–4.3	0.7–2.3
Desert	Cairo	3–5	0–1
	Dubai	3–5	0–1
	Phoenix	3–5	0–1
	Average	3–5	0–1
Tundra	Murmansk	5–10	1–3
	Nuuk	3–5	1–3
	Reykjavik	3–5	3–4
	Average	3.7–6.7	1.7–3.3

(continued)

Table 10.1 (continued)

Biome	City	Temperature increase (annualC)	Precipitation increase (mm/day)
Highland	Quito	3–5	0–1
	Cusco	3–5	0.5–1
	Lhasa	3–5	0–1
	Average	3–5	0.2–1

[a] Based on Hadley Model HadCM3, Hadley Centier of Climatic Prediction and Research
[b]Based on National Science Foundation Climate Change Model

would also be unable to survive under a climate characterized by increased temperature and decreased precipitation. Northern cities, such as Murmansk, Nuuk, Reykjavik, Stockholm, and Moscow, may be able to support a greater diversity of trees as these cities become warmer. In cities where global climate change models have predicted increased precipitation conditions would become favorable to a greater number of species. How the balance between species gained and species lost would work itself out is difficult to predict. Cities experiencing decreased precipitation might have the capacity to irrigate trees in particular environments in order to maintain them. Increased precipitation in some cities, such as Los Angeles, could lead to an increased incidence of root disease from fungal species like *Armillaria mellea* (Dreistadt 1994). This could also be a problem for cities that experience significant decreases in summer precipitation that will have to depend on irrigation to maintain tree cover.

McBride and Lacan (2017) applied the 'substitution of space for time' method (Pickett 1989) to identify street trees in California that would not perform well or survive as the climate became warmer and drier. They identified 48 commonly used street species occurring in 12 of the 16 climate zones of the state that should not be planted due to their lack of adaptations to increasing temperatures and decreasing water supplies.

Warfare had a major impact on urban trees in the twentieth century. There were approximately 100,500 street trees in Tokyo prior to the American firebombing in 1945. The firebombing destroyed 63,000 trees. Due to war time fuel shortages, approximately 10% of the remaining street trees were illegally cut. The population of street trees in Tokyo did not reach per-war levels until the early 1970s (Cheng et al. 1999). The urban forests in other Japanese cities and in cities in Asia and Europe also suffered heavy losses during the Second World War (Cheng and McBride 2006; Stilgenbauer and McBride 2010). A frightening level of urban forest destruction occurred in Hiroshima where a single atomic bomb detonated in 1945. It destroyed thousands of street and park trees in a five square mile area. Hundreds of trees beyond this central blast area were damaged by fire (US Strategic Bombing Survey 1946). International efforts to control the proliferation of nuclear weapons following the Second World War have been only partially successful. Fortunately no atomic weapons have been used to destroy cities after World War II. However, conventional weapons have had devastating effects on urban forests in Korea, Vietnam, Sarajevo, Groznyj, and Baghdad (Lacan and McBride 2009).

An evaluation of the street tree frequency data collected for this study suggests a difference in the dependence upon a single tree species between older and newer cities (Table 10.2). The average frequency of occurrence of the most frequently encountered street trees in older cities was 55.2% in comparison to 36.1% in newer cities. This suggests that older cities are depending more upon a fewer number tree species than is the case in newer cities. The greater use of fewer street tree species in older cities may be the result of having had a longer period of time to select the most dependable tree. London plane tree, for example, has a frequency of occurrence of 67% in both London and Paris, while honey locust, the most frequently encountered street tree in New York City has a frequency of occurrence of only 46%. All three of these cities are in the deciduous forest biome and have a common cultural heritage. London plane is also the top street tree in both Sydney and Johannesburg where its frequencies of occurrence are 21 and 24% respectively. These cities are in different biomes, but have strong cultural and historical ties to London. These ties may account for the preference for London plane tree in both Sydney and London, but the relatively low frequencies of occurrence of the tree in both cities suggests a greater street tree diversity exists in comparison to the older city of London.

The difference in the greater dependency of older cities on fewer street tree species could be an indicator of future urban forest street tree diversity. It may be concluded that over time newer cities, like New York, Johannesburg, Sydney, and Vancouver will depend on fewer species for streets and boulevards. The degree to which this is a good strategy remains to be seen. Paris has moved in this direction because of disease problems associate with horse chestnut. Whether urban monocultures of London plane trees can avoid the type of catastrophe that befell the monocultures of American elm, when the Dutch elm disease arrived in North America, is difficult to predict. Until such a catastrophe occurs one would expect to see an increasing use of dependable species, like London plane, gracing more and more streets in the temperate zone urban forests.

Genetic engineering of new varieties of trees for urban environments has the potential to both increase and decrease tree species in the urban forests of the twenty-first century. Genetic engineering is seen by its supporters as an acceleration of tree breeding and selection programs that have produced widely used cultivar such as London plane (*Platanus x hispanica*), Carolina poplar (*Populus x canadensis*), and Common lime (*Tilia x europaea*). Genetic engineers see solutions to pest problems, new capacities for survival in harsh urban environments, and the creation of more attractive trees through new techniques for recombining genes and gene splicing. Many environmentalists view genetic engineering from a different perspective and voice concerns over contamination of native genotypes and potential ecological disasters. Genetic engineering could potentially produce numerous varieties of trees for urban environments. However, the conservative tendency of urban forest managers to use a limited number of species, which have demonstrated

Table 10.2 Frequency of top street species in old vs. new cities

Biome	Old cities	Top species	Frequency	New cities	Top species	Frequency
Tropical forest	Bangkok	Angsana	32	Singapore	Angsana	31
	Rio de Janeiro	Amendoeira	38			
	Average		35	Average		31
Broadleaf evergreen forest	Tokyo	Gingko	33	Sydney	London Plane	21
				Hong Kong	Indian Laurel Fig	25
	Average		33	Average		23
Deciduous forest	London	London plane	67	New York	Honey Locust	46
	Paris	London plane	67			
	Average		67	Average		46
Coniferous forest	Jacksonville	Live oak	70	Seattle	Basswood	29
				Vancouver	Japanese Flowering Cherry	22
	Average		70	Average		26
Mixed deciduous-coniferous forest	Moscow	Common Lime	83	Beijing	Japanese Pagoda	56
	Stockholm	Common Lime	50			
	Average		66	Average		56
Mediterranean woodland/scrub	Athens	White Mulberry	45	Los Angeles	Indian Laurel Fig	28
				Tel Aviv	Indian Laurel Fig	28
	Average		45	Average		28
Savanna-thorn scrub	Lagos	Umbrella tree	42	Brasilia	Ibirs Pita	41
				New Delhi	Angsana	27
	Average		42	Average		34
Grassland	Tehran	Oriental Plane	54	Buenos Aires	Green Ash	47
				Johannesburg	London Plane	24
	Average		54	Average		36

(continued)

Table 10.2 (continued)

Biome	Old cities	Top species	Frequency	New cities	Top species	Frequency
Desert	Cairo	Indian Laurel Fig	73	Phoenix	Mexican Fan Palm	45
	Dubai	Canary Island Palm	67			
	Average		70	Average		45
Tundra[a]	Murmansk					
	Reykjavik					
	Nuuk					
Highland	Lhasa	Chinese White Poplar	70	Quito	Blackwood Acacia	29
				Cusco	Monterey pine	43
	Average		70	Average		36
Sum			552			361
Average			55.2			36.1

[a]Omitted from comparison because of the small number of species that can be used limit in the Tundra Biome

success in given cities, suggests the engineering of more successful species would in the end lead to reduced diversity in the urban forests of the twenty-first century.

In 2000 the California Pacific Gas and Electric Company and many other electric power utility companies throughout the United States launched the SafeTree program under the motto "The Right Tree in the Right Place". The purpose of this program was to enhance, expand and create greater public awareness about the safety issues concerning proper tree placement near electric power lines. The Pacific Gas and Electric Company dedicated $5 million over a 5 year period to develop an education program aimed at the general public, elementary and secondary public schools, grassroots and community organizations. The program also included grants to communities for tree planting of the "right trees". It is in the interest of electrical power utility companies to promote the planting of trees beneath power lines that will not grow into the power lines and require pruning. It can also be argued that it is in the interest of general public safety to avoid having tree branches contact power lines. Such contacts result in thousands of power outages every year and millions of dollars in property damage due to fires. The approach of the SafeTree program is to dictate the use of a specific list of tree species for planting under and near power lines. The SafeTree program has categorized trees into the following height classes (Minnesota SafeTree Program 2002):

> Low Mature Height (Under 25 feet): Recommended for street tree plantings under electric lines or other areas where growing space above ground is limited. Also recommended for narrow planting lawns (less than four feet wide), planting squares or circles surrounded by concrete, large raised planting containers or other places where underground space for roots will not support large trunk diameters.
>
> Medium Mature Height (30–45 feet): Recommended for planting where the above ground and below ground growing space will allow thirty to forty-five foot mature height trees. Wide planting lawns or medians (four to eight foot wide), large planting squares (eight foot square or greater) and other open areas of at least 50 feet diameter.
>
> Large Mature Height (over 45 feet): Recommended for streets without overhead restrictions where very wide planting lawns or medians (greater than eight feet) will allow for a large root system, trunk diameter and root flare. These large trees are also recommended for parks, meadows or other large open areas where their large size, both above and below ground, will not be restricted or cause damage or liability.

The problem inherent with the recommendations in this classification is the restriction of medium and larger sized to larger planting spaces. The restriction of large trees to "parks, meadows, or other large open areas" would preclude their use in many suburban lots. The attractive older residential areas observed in Jacksonville, Seattle, Phoenix, and Los Angeles are graced with large trees. These trees are often of significant value because of the summer shade they provide to outdoor living space as well as the roofs and walls of houses. They have been maintained in relative safety with proper pruning. The potential change in the structure of the urban forest if SafeTree recommendations are followed could have serious consequences for the function of American urban forests in the twenty-first century.

SafeTree species recommended for planting under power lines also poise a threat to the future diversity of urban forests in the United States. Lists are available for three American cities surveyed in this study: Seattle, Los Angeles, and Jacksonville.

The list for Seattle recommended 32 species for planting along streets under power lines. The chief criterion for the listing of these species was a mature height under 25 feet. Jacobson (1989) in his treatment of the trees of Seattle lists 86 species growing in Seattle that are under 25 feet in height at maturity. The SafeTree list for Los Angeles, and other southern California cities, suggests 18 species in comparison to the 61 species listed by Perry (1989) in his comprehensive work on trees for dry California landscapes. For Jacksonville and adjacent areas of northern and central Florida only six species wee recommended for planting under power lines by SafeTree. Nelson (2003) lists 39 native tree species of Florida commonly used in landscaping that are less than 25 feet in height. Numerous additional exotic tree species, under 25 feet in height at maturity, that could be grown in Jacksonville are described by Rogers and Rogers (1991) and Watkins and Sheehan (1975). The potential narrowing of urban forest species diversity that could result from widespread adoption of the limited lists of trees suggested by SafeTree is of concern to the future diversity of the urban forest.

The mindset that can develop in a program named "SafeTree" with the slogan "The Right Tree in the Right Place" is an additional concern. This term infers there are "unsafe trees", "wrong trees" and "wrong places" for trees. That may be true, but should utility companies that have a vested interest in reducing their operating costs be the arbitrators of tree selection.

Another situation leading to a reduction is urban forest diversity is the problem of allergenic reaction to tree pollen. Two of the cities surveyed in this study were listed among the twenty-five worst cities in the United States to live in due to allergies caused by pollen (Tidwell 2001). Phoenix was ranked number eight and Seattle ranked number 19, primarily due to tree pollen in both cities. Phoenix is among a growing list of cities in the United States that has outlawed the planting of species of *Populus* species because of the allergies caused by their pollen. As allergies caused by tree pollen increase in the future one can expect a further banning of certain tree species.

Common design patterns were prevalent in newer residential suburbs of in Jacksonville, Brasilia, Los Angeles, and Sydney. Newer suburbs in these cities have under grounded utility lines in a utility easement adjacent to the street. These easements were typically 8–10 feet wide and were not used for sidewalks or tree planting. In these suburbs street tree have not been planted by the city or the homeowners on the ground adjacent to the utility easement. Yard trees were planted, but at a much lower density than that typical of street trees in older suburbs. As a result both tree number and tree canopy cover was reduced. Also lacking from these newer suburbs was the cohesiveness that a single street tree species can give a suburban street. Explanations for this pattern had to do with safety and improved aesthetics associated with the under grounding of utilities and with the need to have easy access to utility lines in the future. Unfortunately the tradeoff of tree-lined streets and the ecological and aesthetic values associated with street trees are being lost in these new suburbs.

The under grounding of utilities has also limited the growing space for trees in business districts. Using containers has been a strategy for introducing trees into commercial areas where underground utility lines prevent tree growth. Excellent examples of the use of containers for trees can be seen in New York City and Beijing. The use of containers can be limited where sidewalks are narrow and they may interfere with handicap access. Because of the size limits on trees that can be grown in container they may not contribute as much to total urban canopy cover as trees that are grown in the ground. However, container grown trees are protected from mechanical damage by vehicles and are out of the reach of the de-icing salts and dog urine that cause damage to trees grown in tree pits.

The trend toward athletic parks observed in many cities has created a trend toward the use of fewer trees and tree species. Trees are increasingly regulated to the margins of parks or omitted entirely. It would appear that many of the social activities of young people now take place in shopping malls and the use of parks involves more group athletic activities (Owens 2002). Studies in northern Europe suggest that parks seldom attract teenaged girls except as a place for organized sports. However, trees can and have been used successfully in the design of modern parks to create spaces for teenaged girls, small children, and older adults as well as to delineate soccer fields and other active sports areas (Hood 2002).

A factor leading to the decreased use of trees in inner city parks has been a heighten sense of fear of violent attacks where dense stands of trees obscure views within parks (Gobster 1994). Concerns over safety in urban parks have led to the removal of understory shrubs and smaller trees from groves of trees in many American parks. In many ways the urban forest has become the urban savanna in these inner city parks. Gobster has pointed out a difference in the attitudes about tree density and orderliness in urban parks between people living in the inner city and the suburbs of Chicago. Safety values and a sense of orderliness were ranked high by people living in the inner city, while suburbanites indicated greater preferences of natural appearance and ecological integrity. One can expect this difference in views will affect differences in the designs of future inner city parks in contrast to parks in forested suburbs.

Economic trends apparent at the end of the twentieth century can be expected to continue in the twenty-first century. The last few decades of the twentieth century saw a widening gap between income groups around the world. In 1960 the 20% of the world's people who lived in the richest countries had thirty times the income of the poorest 20%. By 1995 this disparity had increased to 82 times as much income (UNDP 1998). The income of neighborhoods is often reflected in the characteristics of the local urban forest. Schmid (1975) in his seminal study of the Chicago urban forest contrasted tree density in the front yards of houses along an income gradient. Greater tree density occurred in the higher income neighborhoods. Yang et al. (2009) recorded a similar relationship between income level and the amount of tree foliage that could be seen at intersections across income levels in Berkeley, California.

The tree densities in residential areas of three American cities (Jacksonville, Los Angeles, Seattle) were measured as a part of the current examination of the world's urban forest. Middle and three lower income neighborhoods were compared in each city (Table 10.3). An average of 18 trees/sample (0.7 acre) occurred in the lower

Table 10.3 Comparison of tree densities in low and middle income neighborhoods

	Seattle		Los Angeles		Jacksonville		Average	
Location	Low[a]	Middle	Low	Middle	Low	Middle	Low	Middle
Parkway	4[b]	14	2	14	1	7	2	12
Yard	7	39	14	30	12	25	16	31
Total	11	53	16	44	13	32	18	43

[a]Income level of neighborhood
[b]Average number of trees on one side of street per city block in low and middle income residential neighborhoods

income neighborhoods while an average of 43 trees/sample (0.7 acre) occurred in the higher income neighborhoods. The neighborhoods chosen for determining tree density were neither the poorest nor the most expensive neighborhoods in any of the cities, but neighborhoods typical of 15–30 percentiles and the 50–75 percentiles in terms of real estate value. Real estate value was assumed to be an approximate gauge of home owner income. It was apparent in the lower income neighborhoods that little if any disposable income was being spent on tree care or the planting of new trees. In contrast, middle-income neighborhoods showed evidence of tree pruning and recent tree planting. Tree mortality was observed in the lower income neighborhoods of Los Angeles, but not in Seattle and Jacksonville. This mortality may have been induced by a lack of tree watering, a cost lower income families may not be able to afford.

In the middle income neighborhoods of the Los Angeles metropolitan area municipal agencies were caring for street trees while in the lower income neighborhoods no evidence of street tree care was observed. In these middle income residential areas the typical block (approximately 300 feet long) supported 14 street trees. In the lower income neighborhoods an average of only two street trees were present, although there were stumps to suggest greater numbers of trees had occurred along these streets in the past. Lower property values equated to lower tax revenues for cities. In the face of declining urban budgets and increasing management costs many cities have eliminated tree care for street trees. Hazardous trees are removed and fallen trees cleaned up, but few if any street trees are replanted.

The disparity observed in tree density between middle and lower income sections of American cities pales in comparison to that seen in cities in the Third World. In cities like Johannesburg, Quito, Rio de Janeiro, New Delhi, and Cairo the desperately poor live in shanty towns and poor neighborhoods nearly devoid of trees. Although a few newly planted trees were observed in shanty towns Johannesburg and New Delhi their density was extremely low in comparison to the higher income neighborhoods in these cities.

Extrapolation into the future of the differences in tree densities between higher and lower income neighborhoods, in view of an increasing disparity in income, does not present an encouraging picture. This does not necessarily have to be the future if municipal governments recognize the value of the urban forest. A more positive future can be seen in cities like Singapore and Tehran where municipal street tree planting and maintenance programs have been handle in an egalitarian fashion.

There was a downward trend in support for urban forest management in many cities in the North America and northern Europe in the waning decades of the twentieth century. The international economic recession of the 1980s and early 1990s reduced municipal funding with the result that tree care programs were tabled or reduced to a minimum in many cities. Following the 2008 recession in the United States funds for urban tree maintenance were severely limited. In San Francisco the responsibility for tree maintenance and sidewalk repair was shifted to private property owners as a result. Unlike the depression of the 1930s, this more recent economic downturn did not stimulate the creation of Work Progress Administration and Civilian Conservation Corps like projects to assist cities in the management of their urban forests (Cutler 1985).

Support for urban forest management is unlikely to grow in the twenty-first century without strong municipal advocates like Chicago's mayor John Daley, Jr. Mayor Daley campaigned on a promise to improve the urban forests of Chicago, often citing the fact that he was born on Arbor Day. In Los Angeles, mayoral candidate Antonio Villaraigosa promised to plant 1,000,000 trees if elected in 2006. To celebrate his election he planted the first of these trees on the day he was sworn in as mayor. Although the target of 1,000,000 newly planted trees was not reached by the end of Mayor Villaraigosa's term in office (approximately 400,000 trees were planted [AHBELAB 2016]) the tree planting program he initiated is still in operation and trees continue to be planted because of his political leadership.

Non-political advocates for increasing support for urban forest management, like Andy Lipkis in Los Angeles, also help improve the funding picture for urban forestry in the future. Much of their success will depend upon the general public's support for urban forest management. The maintenance of urban forests can be expected to decline in the face of increasing management costs unless citizens in democracies are willing to shoulder additional taxes for improving the urban environment. One factor that may make a significant difference in the public's attitude toward supporting urban forest management activities is the research of Gregory McPherson and his associates with the U.S. Forest Service Urban Forestry Project (McPherson 1992, 1994, McPherson et al. 1999). They have demonstrated in a number of studies the long-term economic benefits of urban forests. The benefit to cost ratio of urban forest management in Chicago was 2.83 and 1.9 in Modesto, California. McPherson's research in Chicago suggested a range in the "payback" period for tree planting from 9 to 18 years depending upon the species and assuming a 7% discount rate. The USFS Urban Forestry Research Project under McPherson's leadership has prepared a number of manuals to assist cities in calculating the benefits-costs ratios of urban forest management. McPherson's work holds promise for increased municipal funding of urban forestry in the United States. The approach has been applied in Beijing by Yang et al. (2005) and hopefully will be applied in other cities around the world to encourage the support of urban forest management.

Although the future funding for urban forest management is unclear at this time, a more definite prediction may be made about actual management activities. Trends in the United States, Singapore and Japan suggest a growing number of volunteers are being used to plant and care for street trees. Training programs like those offered by TreePeople in Los Angeles and Tree Doctors in Tokyo are educating people in tree maintenance that will increase early survival and proper initial pruning of trees. The use of volunteers for tree planting and some maintenance activities in parks has been noted in both Singapore and New York City. Other examples of effective use of volunteers are in the design and maintenance of traffic circles in New Delhi and Seattle. The involvement of volunteers not only reduces the municipal cost of tree maintenance activities, it also builds community support for urban forestry. This support can be important in increasing public support for urban forest management.

Involvement of the public in both urban forest management activities and in the process of urban forest planning can be expected to increase in the twenty-first century. The public's desire to be involved in decisions about their environment includes decisions about where and what kinds of trees should be planted. Increasingly regulations guiding urban planning require citizen review. Although citizen involvement in planning can be time-consuming, and at times contentious, it is in the best interest of the democratic process and in the end will promote citizen support for urban forests. Unfortunate examples of omitting citizen participation in urban forest planning have occurred in Mexico City and Tokyo. In Mexico City an ambitious effort to plant one million trees in preparation for the Olympic in 1964 was thwarted when people pulled thousands of newly planted trees from the ground and threw them into the street. In Tokyo, apartment dwellers have cut down trees that threatened to shade their apartments. These situations could have been avoided with the involvement of people in the urban forest planning process.

The future of the urban forest will be closely tied to the future of the city. Science fiction is probably not the best predictor of future conditions, however, it has on more than one occasion predicted devices, modes of transportation, and urban landscapes that subsequently became commonplace. The nautilus submarine in Jules Verne's "One Thousand Leagues Under the Sea", the rocket ships capable of inter planetary travel in H. G. Wells' "War of the Worlds", and Dick Tracy's two way wristwatch are a few examples. How then does one respond to the future urban landscape portrayed in the 1984 American movie "Blade Runner" and the 1996 French film "The City of Lost Children". Los Angeles in "Blade Runner" sits in the same topography as the city today, but it is a city without trees. Neither in the aerial views of the city nor in the street level scenes are any trees in evidence. The same is true for the futuristic French city portrayed in the "The City of Lost Children". These are grim prospects of potential urban landscapes.

Near the middle of the twentieth century English and American authors were predicting the unraveling and demise of the city from both a physical and sociological standpoint (Orwell 1949; Jacobs 1961). George Orwell predicted in his novel "1984" an urban structure where the "undesirables" were walled out of those portions of the city where behavioral standards of the ruling class are enforced. The wastelands occupied by the "undesirables" are the crumbling remains of the former city where municipal maintenance and authority have been abandoned. Jane Jacobs predicted decay in the quality of life in the city due to planning professional's lack of understanding of urban life. The breakdown in urban diversity caused by the replacement of older neighborhoods with high-rise apartment buildings, the separation of primary types of land use (housing, work, recreation), and the dissection of functional neighborhoods by freeways would, in Jacobs view, lead to the death of American cities.

Joel Garreau (1991) in a book entitled "Edge City" explored the future of the American metropolitan landscape. He saw future urban development as clusters of high rise centers for business and employment emerging at intersections of interstate freeways and rig roads surrounding existing cities. Suburban housing developments will evolve around these new centers to provide housing for the workforce employed in the centers. Garreau suggests examples of this future could be seen in "edge cities" that were developed around Manhattan, Boston, Atlanta, Dallas, San Francisco and other U.S. cities. The vision of the future suggested by the "Edge City" is one in which trees will play a limited role as long as they do not minimize a sense of safety, upscale ambience, and incur unnecessary maintenance costs. Garreau introduced the derogatory term "tree shit" in reference to the by products of the life cycle of trees that fall to the ground. He suggests that the cleaning up of such debris is seen as the primary argument for not putting trees in parking lots. This attitude, if shared by future developers, does not speak well for the future of the urban forest in the edge city.

Unfortunately, the futuristic visions of Orwell, Jacobs, and Garreau are beginning to come true. Edge cities continue to emerge and their adjacent gated communities now separate different economic and social classes in many of the world's cities. The vast parking lots common to edge cities support few if any trees. Tree canopy cover in the high rise sections of edge cities is very limited. The contrasts of urban forest conditions inside and outside of the gated communities is associated with edge cities striking. The affluence of the residents of gated communities supports the planting and care of street and yard trees, while in less affluent neighborhoods tree canopy cover and tree care continue to diminish.

The conversion of older neighborhoods into residential estates dominated by high rise condominiums and apartment buildings continues both in American cities as well as in other parts of the world. In Beijing the most tree-species rich neighborhoods were the old hutongs where, dozens of fruit, nut, and medicinal species were reported in 1992 by George Profous. Most of these former hutongs have now been leveled and replaced by high rise apartment buildings. Usually only a few tree species have been planted adjacent to these new buildings. Not only has species diversity been reduced by the conversion of hutongs to apartment buildings ands condominiums, but also the tree canopy cover has often been reduced by more than 70%.

Jacobs renewed her concern for the future of urban life in her book "Dark Age Ahead" (Jacobs 2004). In this book she identifies five pillars of our culture that are in serious decline. One of these, the effective practice of science, can have consequences for the future of urban forests. Unless the municipal governments apply the best scientific knowledge to problems of urban forest management one can expect a general decline in the health, and subsequently, the structure and function of the urban forest. It is of course imperative that urban forest research effectively address new issues in urban forest management as well as those issues that have yet to be completely resolved. High priority research for the twenty-first century should include studies of the (1) genetic nature of trees that do well in urban environments, (2) nature of soil amendments to improve urban soils and engineered soils, (3) refinement of benefits-costs analysis, (4) relationship between diversity and stability in the urban forest, (5) new approaches to confront insects and disease problems, (6) understanding the social and psychological values of trees, and (7) effective methods for community participation in urban forest planning and management.

The composition, design, function, and management of future urban forests in developing nations must respond to the needs of increasing populations of urban poor, many of whom will be new migrants to the city in the twentieth century. Their needs for firewood, building material, fodder, and food can be partly served by urban forests. These forests must be planned to address these needs in addition to the aesthetic, psychological, and ecological functions normally associated with urban forests. Kuchelmeister and Braatz (1993) emphasize the necessity of local planning to address the particular needs of resource-poor people. The need for firewood may be particularly acute in one area, while the improvement of nutritional status more demanding in another. The resources needed by people in these developing countries will most likely be provided by peri-urban forests. A significant issue in the establishment of these peri-urban forests will be the acquisition of land. Some existing land within cities can serve to address some of the needs of resource-poor people. The examples of fruit tree planting along boulevards and in city parks in Moscow, on vacant lots in Tehran, and on school, church, and service organization properties in Los Angeles suggest a potential for some production within the bounds of the traditional urban forest. Equal to the task of the establishment of these resource-providing units of urban forests will be their management as commons for the good for the greatest number of resource-poor urban dwellers.

From the above discussion it can be concluded that the urban forests of the world may have more than one future. Variations in needs and motivations of urban dwellers will differ in the world's cities. Economics and ecology will also account for cycles that will affect the composition, design, function, and management of our urban forests. What one does not see changing is the basic admiration people have for trees (Jonnes 2016). This admiration may be put to a test during the twenty-first century, but in the long run it will be the key element in sustaining the urban forest.

The evaluations of about one-half of the factors discussed in the above paragraphs suggest a declining future for the urban forest, while the other half present a more positive picture. A balance among these factors will determine the characteristics of the urban forests in the world's cities in the twenty-first century. If the balance is dominated by positive factors, like volunteer involvement in urban forest management and a greater understanding of the benefits-costs ratios of urban forests, then the urban forests of the twenty-first century can be expected to take on even a greater character than was imagined in the early twentieth century. War, continued decline in municipal budgets, natural disasters, and pest problems could tip the balance toward forests characterized by less diverse species composition, structure, and function. However, it does not seem likely that cities will be deprived of trees in the twenty-first century, as many cities were in the early middle ages. The desire of people to have trees in their urban environments is a fundamental desire. It can be seen in people the planting of trees in the poorest shanty towns as well as in the richest estates. This basic aspect of human nature will sustain urban forests in all of the world's biomes.

Literature Cited

Aas G, Riedmiller A (1994) Trees of Britain and Europe (trans: Walters M). HarperCollins, London
Adams RM (1960) The origin of cities. Sci Am 171:153–168
AHBELAB (2016) L.A.'s million trees… maybe a little less. https://ahbelab.com/2016/03/03/l-as-million-trees-maybe-a-little-less/
Alberti L (1485) De re aedificatoria (completed 1452; first published 1485)
Alphand A (1868–73) Les promemdades de Paris. J. Rothschild, Paris
Anonymous (2010) Stormwater design guidelines. San Francisco, CA. www.sfwater.org/.../show-document.aspx?...
Antwerp City Archives (1578) Order for planting old city wall. PK 917, fol.163, 18 December 1578.
Appleyard D et al (1981) Livable streets. University of California Press, Berkeley
Arnold HF (1993) Trees in urban design. Van Nostrand Reinhold, New York
Associated Press (2000) Heavy rains claim 15 lives in Ecuador capital. Associated Press, New York
Aurousseau M (1928) The distribution of populations: a constructive problem. Geogr Rev 11
Bacon F (1625) The essays. Penguin Books, New York, 1985
Bailey RG (1996) Ecosystem geography. Springer, New York
Bancroft HH (1889) History of Arizona and New Mexico, 1530–1885. University of California Press, Berkeley
Barnard ES (2002) New York City trees. Columbia University Press, New York
Bartholmew J (1950) The advanced atlas of modern geography. Meiklejohn Sons, London
Barton N (1962) The lost rivers of London. Historical Publications Ltd, London
Beatty RA, Hickman CT (1981) Survey of urban tree programs in the United States. Urban Ecol 5(2):81–102
Beatty RA (1999) (Personal communication)
Beijing Annals Editorial Board (2000) Landscape plants. In: Beijing Annals Editorial Board (ed) Beijing Annal, Vol. 16 Municipal Administration, copy 52, Garden and Afforestation Annals. Beijing Publishing House, Beijing, pp 337–384
Benjamin MT, Winer AM (1998) Estimating the ozone-forming potential of trees and shrubs. Atmos Environ 32:53–68
Ben-Joseph E (1995) Creative flows: the Japanese experience in urban stream restoration. On The Ground 2(1):19–20
Beveridge CE, Schuyler D (1982) In: McLaughlin C, Beveridge CE (eds) Creating Central Park, 1857–1861. The papers of Frederick Law Olmsted. John Hopkins University Press, Baltimore
Bioswale (2012) Bioswale Garden. Cornell University. College of Agricultural and Life Sciences. http://www.cornellplantations.org/our-gardens/botanical/bioswalegarden

J.R. McBride, *The World's Urban Forests*, Future City 8,
DOI 10.1007/978-3-319-52108-4

Bossard CC et al (eds) (2000) Invasive plants of California's wildlands. University of California Press, Berkeley
Bradshaw AD (1976) Derelict land – is tiding going to far? Planner 65:85–88
Breckle S (2002) Walter's vegetation of the earth: the ecological systems of the geo-biosphere, 4th edn. Springer, Berlin. (Translated from the 7th, completely revised and enlarged German edition by Gudrun and David Lawlor)
Brittin R (2003) New study reveals dramatic national tree loss. http://www.americanforests.org/news/print.php?id=120
Brown PRL (1988) The body and society. Columbia University Press, New York
Buchmann A (2002) Lucio Costa, O inventor da cidade de Brasilia. Thesaurus, Brasilia
Burchardt J (1877) Die Kultur der Renaissance in Italien (English translation by Middlemore SGC, 3rd edition 1951). Harper and Brothers, New York
Burke GL (1960) The making of Dutch towns: a study in urban development from the tenth to the seventeenth centuries. Simmon-Broardman, New York
Campanella TJ (2003) Republic of shade: New England and the American elm. Yale University Press, New Haven
Cardelino CA, Chameides WL (1990) Natural hydrocarbons, urbanization, and urban ozone. J Geophys Res 95(D9):13971–13979
Carroll M (2003) Earthly paradises: ancient gardens in history and archaeology. British Museum Press, London
Carroll-Spillecke M (1992) The gardens of Greece from Homeric to Roman times. J Gard Hist 12(2):84–101
Catlin G (1837–39) Bird's eye view of the Mandan Village, 1800 miles above St. Louis. Smithsonian American Art Museum. In: G Gurney, Heyman TTT (eds) 2002 George Catlin and his Indian Gallery. W. W. Norton and Company, New York
Chadwick G (1966) The park and the town: public landscape in the nineteenth and twentieth centuries. Praeger, New York
Chameides WL et al (1988) The role of biogenic hydrocarbons in urban photochemical smog: Atlanta as a case study. Science 241:1473–1475
Chang K (1986) The archaeology of China. Yale University Press, New Haven
Chapela I et al (2001) Ectomycorrhizal fungi introduced with exotic pine plantations induce soil Carbon depletion. Soil Biol Biochem 33(12–13):1733–1740
Chen ZX (1998) Research of the ecological benefits of green space in Beijing city. Chin Landsc Gard 14:57–60
Cheng S, McBride JR (2006) Restoration of the urban forests of Tokyo and Hiroshima following World War II. Urban For Urban Green 5:155–168
Cheng S, McBride JR, Fukunari K (1999) The urban forest of Tokyo. Arboricult J 23(4):379–392
Cheng S, McBride JR, Fukunari K (2000) The urban forest of Tokyo. Arboricult J 23:379–392
Clark GT (1884) Medieval military architecture in England. Wyman & Sons, London
Clifford D (1963) A history of garden design. Praeger, New York
Clough AH (ed) (2001) Plutarch's lives. Random House, New York
Committee on Public Walks (1833) Report of the select committee on public walks. House of Commons, London
Conard SG et al (1980) Riparian vegetation and flora of the Sacramento Valley. In: Sands A (ed) Riparian forests of California. Their ecology and conservation. University of California, Division of Agricultural Sciences, Davis, pp 47–57
Cooper Marcus C (1995) House as a mirror of self: exploring the deeper meaning of home. Conrai Press, Berkeley
Cowell FR (1978) The garden as a fine art from antiquity to modern times. Houghton Mifflin, Boston
Crisp F (1924) Medieval gardens. Brentano's, New York. (2 vols)
Crouch DP et al (1982) Spanish city planning in North America. MIT Press, Cambridge, MA
Cutler P (1985) The public landscape of the new deal. Yale University Press, New Haven

Dansereau P (1957) Biogeography: an ecological perspective. The Ronald Press Company, New York
Davis MG, Zabinski C (1992) Changes in geographic range resulting from greenhouse warming: effects on biodiversity in forests. In: Peters R, Lovejoy T (eds) Global warming and biodiversity. Yale University Press, New Haven
de Boer G (1973) The two earliest maps of Hull. Post Medieval Archaeol 7:79
de Candolle AP (1820) Essai elementaire de geographie botanique. F. Levrault, Paris
de Norri B et al (1986) Accidentes Climatios y Gestion de las Quebradas de Quito. In: Paisajes Geograficos, vol 17. CEPEIGE, Quito, pp 25–44
Delcourt HR, Delcourt PA (2000) Eastern deciduous forests. In: Barbour MG, Billings WD (eds) North American terrestrial vegetation. Cambridge University Press, Cambridge, pp 358–395
Department of Greenspace (1998) Greeenspace, Section 13. Department of Greenspace, City of Tehran
Diamond J (1997) Guns, germs, and steel. W. W. Norton and Company, New York
Diamond J (2004) Collapse: how societies choose to fail or succeed. Reinhold, New York
Diaz B (1632) Historia verdadera de la conquista de la Nueva Espana. Emprenta del Reyno, Madrid
Dixon DM (1969) The transplantation of punt incense trees in Egypt. J Egypt Archaeol 55:55–65
Downing AJ (1852a) Shade trees in cities. Horticulturalist 8:346–347
Downing AJ (1852b) Landscape gardening and rural architecture. Dover, New York (reprint, 1991)
Dreistadt S (1994) Pests of landscape trees and shrubs, Publ. No. 3359. University of California Division of Agriculture and natural resources, Oakland
Duryea ML, Kampf E, Littell RC (2007a) Hurricanes and the urban forest: I. Effects on Southeastern U.S. coastal plain tree species. Arboricult Urban For 33(2):83–97
Duryea ML, Kampf E, Littell RC, Rodríguez-Pedraza CD (2007b) Hurricanes and the urban forest: II. Effects on tropical and subtropical tree species. Arboric Urban For 33(2):98–112
Du Val T (2003) Personal communication about roof gardens in New York
Dwyer JF et al (1994) The deeper significance of urban trees and forests. In: Platt RH et al (eds) The ecological city. University of Massachusetts Press, Amherst, pp 137–150
Dziobeck E (1992) Das Grab des Ineni. Theben Nr. 81. Mainz
Eco U (1986) Art and beauty in the middle ages. Yale University Press, New Haven
Ellenberg H, Mueller-Dombois D (1973) International classification and mapping of vegetation. UNESCO ecology and conservation series No. 6. UNESCO, Paris. 93p
Evelyn J (1641) Diary entry. October 5, 1641
Everard LC (1926) Arbor Day: its purpose and observance, Farmers bulletin No 1492. U.S. Dept. of Agriculture, Washington, DC
Eyre SR (1968) Vegetation and soils: a world picture. Edward Arnold, London
Fabos JG et al (1968) Frederick Law Olmsted, Sr: founder of landscape architecture in America. University of Massachusetts Press, Amherst
Fairgrieve J (1921) Geography and world power. University of London Press, London
Faulkner W (1936) Absalom, Absalom! Scribner, New York
Fein A et al (1972) Frederick law Olmsted and the American environmental tradition. Braziller, New York
Felt EP (1939) Hurricane damage to shade trees. Am For 45(1):20
Ferguson W (1948) The renaissance in historical thought: five centuries of interpretation. Houghton Mifflin, Boston
Finch VC, Trewartha GT (1936) Elements of geography. McGraw-Hill, New York
Finch VC, Trewartha GT (1949) Elements of geography. McGraw-Hill, New York
Fong YK, Lim J (1999) The wilt of our beautiful Angsana. Gardenwise. The Newsletter of the Singapore Botanic Gardens, vol. XII. July 1999, ISSN 12-1688. 20 pp
Fox T et al (1985) Struggle for space: the greening of New York. Neighborhood Open Space Coalition, New York
France RL (2002) Handbook od water sensitive planning and design. CRC Press, Boca Raton

Francis M et al (1994) Community open spaces: greening neighborhoods through community action and land conservation. Island Press, Covelo
Fraser PM, Nicolas B (1958) The funerary garden of Mousa. J Roman Stud 48:117–129
French S (1975) The cemetery as cultural institution: the establishment of Mount Auburn and the 'Rural Cemetery' movement. In: Stannard DE (ed) Death in America. University of Pennsylvania Press, Philadelphia
Fricker L (1974) John Claudius Loudon: the plane truth? In: Willis P (ed) Furor hortensis: essay on the history of the English landscape garden in memory of H F Clark. Elysium Press, Edinburgh
Fries SD (1977) The urban idea in colonial America. Temple University Press, Philadelphia
Gakenheimer R (1959) The Spanish King and his continent: a study of the importance of "The Laws of the Indies" for urban development in Spanish America. Master's thesis, Cornell University
Gangotena et al (1990) Tree species along major arterial and residential streets of Quito. Department of Parks and Gardens, Quito (in Spanish)
Garreau J (1991) Edge city. Anchor Books, New York
Geron CD et al (1994) An improved model for estimating emissions of volatile organic compounds from forests in the eastern United States. J Geophys Res 99(D6):12773–12791
Ghannam J (ed) (1997) Exotic Chinese insect. The Forestry Source. March, 1997
Girouard M (1985) Cities and people: a social and architectural history. Yale University Press, New Haven
Gleason KL (1994) Ponticus Pompeiana: a new perspective on the first public park of ancient Rome. J Gard Hist 14(1):13–27
Gobster PH (1994) The urban savanna. Restoration Manag Notes 12(1):64–71
Gravan ABN (1963) Proprietary Philadelphia as artifact. In: Handlin O, Burchard J (eds) The historian and the city. M.I.T. Press, Cambridge, MA
Gray DH, Leiser AT (1982) Biotechnical slope protection and erosion control. Krieger Publishing Company, Malabar
Greater Vancouver Sewerage & Drainage District (2012) Stormwater source control design guidelines 2012. Lanrc Consultants Ltd., Keer Wood Leidal Associates, Ltd. and Goya Ngan. Metro Vancouver
Greenery Research Laboratory (1981) Report on the technology of roadside greenery. Civil Engineering Research Institute, Tokyo
Grey GW (1996) The urban forest: comprehensive management. John Wiley, NY
Grey GW, Deneke FJ (1978) Urban forestry. Wiley, New York
Griffith FL (1924) Excavations at El-Amarnah. J Egypt Archaeol 10:299–305
Gruffydd B (1987) Tree form, size, and colour: a guide to selection, planting, and design. E. & F. N. Spon, New York
Guichard S, Augustin S (2002) Acute spread in France of an invasive pest, the horse chestnut leaf miner *Cameraria ohridella* Desehka & Dimic (Lep, Gracillareidae). Anzeiger fur Schadlingskunde 75(6):145
Hagen BW (2004) New pests threaten urban eucalyptus. Public Works City of Manhattan Beach, CA. Jan. 15, 2004
Hall P (1998) Cities and civilization. Pantheon Books, New York
Hammond CS (1961) Hammond's ambassador world atlas. C. S. Hammond and Company, Maplewood
Harris RW et al (1999) Arboriculture, 3rd edn. Prentice Hall, Upper Saddle River
Harmon PF, Harmon CL (2004) Sudden oak death. IFAS Extension. University of Florida. Gainsville. http://www.stoppinginvasives.com/dotAsset/743fe3fa-21ed-4e0f-a070-cbe7325bdc80.pdf
Harvey JH (1989) Garden plants of around 1525, The Fromond list. Gard Hist 17(2):122–134
Haus-Ruker Inc. (1976) Rooftop oasis project: co-op owner's and landlord's guide to organizing a rooftop project. Haus-Rucker, Inc., New York
Hay J (1973) Ancient China. Bodley Head, London

Heikamp D (1967) La Villa di Castello. L'Oeil 2, nos. 151–153, pp 56–63
Helton T (ed) (1964) The renaissance: a reconsideration of the theories and interpretations of the age. University of Wisconsin Press, Madison
Hepper FN (1990) Pharaoh' flowers: the botanical treasures of Tutankhamen. HMSO, London
Hien WN et al (2002) Field measurements of thermal parameters of rooftop gardens. Handbook on skyrise greening in Singapore. National Parks Board and National University of Singapore, Singapore
Hiorns FR (1956) Town-building in history. G. G. Harrap, London
Hodel DR (1988) Exceptional trees of Los Angeles. California Arboretum Foundation, Los Angeles
Hodel DR (2005) Species composition of the Los Angeles urban forest. Personal communication
Hoffman U (1976) Problems of the city climate of Stuttgard. In: Franke E (ed) City climate: data and aspects of city planning. Stutgard, Karl Kramer
Holmes FW (1978) Dutch elm disease: perspectives after sixty years. Northeast Regional Research Publication, Cornwell University, Ithaca
Hood W (2002) The design for Lafayette Park, Oakland, CA. Personal Communication
Horn W, Born E (1979) The plan for St. Gall. University of California Press, Berkeley
Hough M (1995) Cities and natural process. Routledge, New York
http://www.americanforest.org/resources/urbanforests/treedeficit.php
http://www.americanforests.org/news/print.php?id=120
http://www.centralparknyc.org Central Park Conservancy
http://www.treepeople.org TreePeople
https://www.itreetools.org/about.php i-Tree Tools
Huntington E (1924) Civilization and climate. Yale University Press, New Haven
IMQ (1992) Illustre Municipio de Quito, Direccion de Planificacion. Quito en Cifras. IMQ, Quito
Ingersoll R (1995) Medieval cities: Burges and Florence. Lecture presented at Rice University 10/9/95. http://www.owlnet.rice.edu/~arch343/lecture9.html
Irving RG (1981) Indian summer: Lutyens, Baker and imperial Delhi. Yale University Press, New Haven
Jackson JB (1970) Several American landscapes. In: Zube EH (ed) Landscapes: selected writings of J B Jackson. University of Massachusetts, Amherst
Jackson JB (1984) Discovering the vernacular landscape. Yale University Press, New Haven
Jackson KT, Vergara CJ (1989) Silent cities: the evolution of American cemeteries. Princeton Architectural Press, New York
Jacobs AB (1993) Great streets. MIT Press, Cambridge, MA
Jacobs J (1961) The death and life of the great American cities. Random House, New York
Jacobs J (2004) Dark age ahead. Random House, New York
Jacobs AB, MacDonald E, Rofe Y (2002) The boulevard book. The MIT Press, Cambridge, MA
Jacobson AL (1989) Trees of Seattle. Sasquatch Books, Seattle
Jacobson AL (1996) North American landscape trees. Ten Speed Press, Berkeley
Jashemski WF (1979) The gardens of Pompeii, Herculaneum and the Villas destroyed by Vesuvius. Caratzas Brothers, New Rochelle
Jashemski WF (1981) The campanian peristyle garden. In: Macdougall EB, Jashemski WF (eds) Ancient Roman gardens. Dumbarton Oaks Colloquium on the history of landscape architecture VII. Washington, DC
Jim CY (1996) Roadside trees in urban Hong Kong: Part I. Census methodology. Arboricult J 20:221–237
Jim CY (1999) A planning strategy to augment the diversity and biomass of roadside trees in urban Hong Kong. Landsc Urban Plan 44:13–32
Johnson L (2016) SF supes' plans to return street tree maintenance to city. San Francisco Chronicle. July 19, 2016. http://www.sfchronicle.com/politics/article/SF-supes-plans-return-street-tree-maintenance-8385317.php
Jolley HE (1989) Parkways. In: Tishler WH (ed) American landscape architecture. Ephrata

Jones WHS (translator) (1935) Pausanias: description of Greece. William-Heinemann, London

Jonquiere P (1978) Woonerf: an environment for man and transport together–the present aspect. Second international symposium on man and transport – the future aspects. Tokyo, Japan

Jonnes J (2016) Urban forests: a natural history of trees and people in the American cityscape. Viking, New York

Josephy AM Jr (1994) Five hundred nations. Knopf, New York

Josselyn J (1663) Account of two voyages to New England. Massachusetts Historical Society, Collections, 3rd ser. (Cambridge, 1838), 3:318, 321

Jurries D (2003) Biofilters. State of Oregon Department of Environmental Quality. Salem. http://www.deq.state.or.us/nwr/stormwater.htm

Kaufman W (1967) Stuttgard cleans the air. Landsc Architect 57:176

Kazakov L (1999) Trees growing in Murmansk. Polar Arctic Botanical Garden, Apatity

Kemp BJ (1989) Ancient Egypt: anatomy of a civilization. Routledge, London

Kielbaso JJ (1990) Trends and issues in city forests. J Arboric 16(3):69–76

Kittredge J (1948) Forest influences: the effects of woody vegetation on climate, water, and soil with applications to the conservation of water and the control of floods and erosion. Dover, New York

Kollin C (1991) On balance: weighing the benefits and costs of urban trees. USFS. Northeastern Forest Experiment Station. Urban Forestry Project, Syracuse

Köppen W (1931) Grundriss der Klimakunde. Walter De Gruyter and Company, Berlin

Kostof S (1991) The city shaped: urban patterns and meanings through history. Bullfinch Press, New York

Kostof S et al (1979) History of urban form before the industrial revolution. George Godwin, Ltd., London

Kovacs K, Vaclavik T, Haight RG, Pang A, Cunniffe NJ, Gilligan CA, Meentemeyer RK (2011) Predicting the economic cost and property value loss attributed to sudden oak death damage in California (2010–2020). J Environ Manag 92(4):1292–1302

Krussmann G (1984) Manual of cultivated broad-leaved trees and shrubs. Timber Press, Beaverton

Krussmann G (1985) Manual of cultivated conifers. Timber Press, Portland

Kuchelmeister G, Braatz S (1993) Urban forestry revisited. Unasylva 44(173):3–12

Kuchler AW (1967) Vegetation mapping. Ronald Press, New York

Kuchler AW (1973) Problems in classifying and mapping vegetation for ecological regionalization. Ecology 54:512–523

Lacan I, McBride JR (2009) War and trees: the destruction and replanting of the urban and peri-urban forests of Sarajevo, Bosnia-Herzegovina. Urban For Urban Green 8:315–323

Landsberg S (1995) The medieval garden. Thames and Hudson, New York

Landsberg S (1996) Bayleaf – a medieval yeoman's garden. The Garden 121:6

Lavedan P (1975) Historie de l'urbanisme a Paris. Hachette, Paris

Lawrence HW (1988) Origins of the tree-lined boulevard. Geogr Rev 78(4):355–374

Lawrence HW (1993) The greening of the squares of London: transformation of urban landscapes and ideas. Ann Assoc Am Geogr 83(1):90–118

Lawrence HW (1995) Changing forms and persistent values: historical perspectives on the urban forest. In: Bradley GA (ed) Urban forest landscapes: integrating multidisciplinary perspectives. University of Washington Press, Seattle, pp 17–40

Lawson L (1997) Urban agriculture and its design implications. Ph.D. dissertation. University of California, Berkeley

Lazzaro C (1990) The Italian renaissance garden. Yale University Press, New Haven

Lewington A (1990) Plants for People. The Natural History Museum, London

Lewis HT (1973) Patterns of Indian burning in California: ecology and ethnohistory, Anthropological papers No. 1. Ballena Press, Ramona

Liebhold AM et al (1995) Invasion by exotic forest pests: a threat to forest ecosystems. For. Sci. Monogr. 30. 49 p

Linden-Ward B (1989) Cemeteries. In: Tishler WH (ed) American landscape architecture. The Preservation Press, Washington, DC
Lipkis A, Lipkis K (1990) The simple act of planting a tree. J.P. Tarcher, Los Angeles
Littlewood M (1988) Tree detailing. Butterworth Architecture, London
Litton KB Jr., Tetlow RJ (1978) A landscape inventory framework. Research paper PSW-135. USDA Forest Service, Berkeley. 84 p
Liu XY, Ren R (2003) Characteristics and formation cause of air pollution in Beijing city. City Disaster Reduct 1:41–43
Lloyd CL (2004) Dream house spoiled by tree preservation law. Real Estate News. http://www.pennysaverusa.com/realestate/realnews/story
Lorants S (1946) The new world: the first pictures of America. Duell, Sloan & Pearce, New York
Lousley JE (1944) The pioneer flora of bombed sites in Central London. Rep Bot Exch Club 1941(42):528–531
Louv R (2005) Last child in the woods. Workman Publishing Company, New York
Martin AC et al (1951) American wildlife plants. McGraw-Hill, New York
Massachusetts Bay Records (1861) Essay on the laying out of towns, & c, (1:30). Massachusetts Historical Society, Collections, 5th series (Boston, 1861), 1:480
McBride JR (2003) Palo Alto tree study. Report to the City of Palo Alto Planning Department.
McBride JR, Doughovnikoff V (2011) Characteristics of the urban forests in arctic and near-arctic cities. Urban For Urban Green 11(2):1–7
McBride JR, Froelich D (1984) Structure and condition of older stands in parks and open space areas of San Francisco, CA. Urban Ecol 8:165–178
McBride JR, Gerhard D (1995) Urban riparian woodland ecology in the San Francisco Bay Area. In: Proceedings of the seventh national urban forestry conference. New York City, 11–16 September, 1995. American Forestry Association. Washington, DC, pp 236–239
McBride JR, Jacobs D (1986) Presettlement forest structure as a factor in urban forest development. Urban Ecol 10:26–52
McBride JR, Lacan I (2017) The impact of climate change of street trees in California. Urban For Urban Green (in press)
McBride JR, Mossadegh A (2000) Tree-lined canals and the urban forest of Tehran. Arboricult J 24:155–173
McBride JR et al (2001a) Urban forest effects on air pollution in Hefei, People's Republic of China. Technical report. Hefei Urban Forest Department. Hefei. 32 p
McBride JR et al (2001b) Urban forest effects on air pollution in Ningbo, People's Republic of China. Ningbo Urban Forest Department. Ningbo, 32 p
McDonnell MJ et al (1997) Ecosystem processes along an urban-to-rural gradient. Urban Ecosyst 1(1):21–36
McPherson EG (1992) Accounting for benefits and costs of urban greenspace. Landsc Urban Plan 22(1):41–51
McPherson EG (1994) Modeling benefits and costs of tree planting and care in Chicago. In: McPherson EG et al (eds) Chicago's urban forest ecosystem: results of the Chicago urban forest climate project, General technical report NE-186. USDA Forest Service. Northeastern Forest Experiment Station, Radnor, pp 3–18
McPherson EG, Johnson CW (1988) A community forestry planning process: case study of citizen participation. Landsc Urban Ecol 15:185–194
McPherson EG et al (1999) Benefits and costs of Modesto's municipal urban forest. J Arboric 25(5):235–248
McPherson EG et al (2003) Expanding street tree canopy cover and repairing sidewalks in the City of Los Angeles. Western Arborist 29(3):1–4
Michaels A (1997) Fugitive pieces. Knopf Press, New York
Michelet J (1855) History of France. In: Lausanne (ed) The renaissance, vol 7. Recontre
Miller NF (1978) Excavations at Nemea 1977. Hesperia 47:58–88

Miller W (1997) Urban forestry: planning and managing urban greenspaces. Prentice Hall, Upper Saddle River
Miller RW, Hauer RJ, Werner LP (2015) Urban forestry: planning and managing urban greenspace. Waveland Press, LongGrove
Minnesota SafeTree Program (2002) www.safetree.net
Mitchell AL, House S (1999) David Douglas explorer and botanist. Aurum Press, London
Mobius K (1877) Die Auster send die Austernwirtschaft. Verlag von Wiegandt, Hemple and Parey, Berlin. (Trans. Rice HJ and published by Fish and Fisheries Annual Report of the Commission for the year 1880, vol 3, no. 29, app. H, pp 681–747)
Moll G (1997) America's urban forests: growing concerns. American Forest (August, 1997)
Montero MI (2001) Roberto Burle Marx: the lyrical Landscape. University of California Press, Berkeley
Mooney H (ed) (1977) Convergent evolution in Chile and California: Mediterranean climate ecosystems. Dowden, Hutchinson and Ross, Stroudsburg
Moore A et al (1984) Plan de Manejo Bosque Protector Pinchincha. AID PRONAF, Quito
Mossadegh A (2004) World forest geography. University of Tehran, Karajh
Mosser M, Teyssot G (1991) The architecture of western gardens. MIT Press, Cambridge, MA
Mozingo L (1997) The aesthetics of ecological design: seeing science as culture. Landsc J 16(1):46–59
Mumford L (1961) The city in history. Harcourt Brace Jovanovich, Inc., New York
Murray S (1999) "Human Nature". The shaping of the urban ecosystem in spontaneous settlements of Quito, Ecuador. Ph.D. dissertation. University of California, Berkeley
Nagasaki S (1998) Sensai no ato tazunete – Tokyo o aruku (Searching the traces of war – walk in Tokyo). Agane Gijutsu Senta, Tokyo. 158p
National Association of City Transportation Officials (2013) Urban street design guide. Island Press, Covelo
Nelson GL (2003) Florida's best native landscape plants. University of Florida Press, Gainesville
New England Farmer 12 (October 9, 1833):101.
Newman LA et al (1999) Remediation of trichloroethylene in an artificial aquifer with trees: a controlled field study. Environ Sci Technol 33(14):2257–2265
Newton NT (1971) Design on the land: the development of landscape architecture. Harvard University Press, Cambridge, MA
Nicholson PH (1981) Fire and the Australian aborigine – an Enigma. In: Gill AM, Groves RH, Noble IR (eds) Fire and the Australian biota. Australian Academic Press, Canberra, pp 55–76
Nowak DJ (1994a) Air pollution removal by Chicago's urban forest. In: McPherson EG, Nowak DJ, Rowntree RA (eds) Chicago's urban forest ecosystem: results of the Chicago urban forest climate project, General technical report NE-186. USDA Forest Service, NEFES, Radnor, pp 63–82
Nowak DJ (1994b) Urban forest structure: the state of Chicago's urban forest. In: McPherson EG et al (eds) Chicago's urban forest ecosystem: results of the Chicago urban forest climate project, General technical report NE-186. U.S. Department of Agriculture, Forest Service, Northeastern Forest Experiment Station, Radnor, pp 3–18
Nowak D (2005) Estimates of isoprene and monoterpene production by trees in New York City. Personal communication
Nowak DJ, Crane DE (2000) The urban forest effects (UFORE) model: quantifying urban forest structure and functions. In: Hansen M, Burk T (eds) Integrated tools for natural resources inventories in the twentieth century: proceedings of the IUFRO conference; 16–20 August 1998, Boise. General technical report NC-212. U.S. Department of Agriculture, Forest Service, North Central Forest Experiment Station, St. Paul, pp 714–720
Nowak DJ, Crane DE (2006) The urban forest effects (UFORE) model: quantifying urban forest structure and functions. In: Hansen M (ed) Second international symposium: integrated tools for natural resources inventories in the twentieth century. General technical report. USDA Forest Service

Nowak DJ et al (2000) A modeling study of the impact of urban trees on ozone. Atmos Environ 34:1601–1603
Nowak DJ et al (2003) The urban forest effects (UFORE) model: field data collection manual. USDA Forest Service, Northeastern Research Station, 5 Moon Library, SUNY-ESF, Syracuse
Nuttall Z (1921) Royal ordinances concerning the laying out of new towns. The Hispanic Am Hist Rev 4(Nov):743–753
Orwell G (1949) Nineteen-eighty-four: a novel. Signet Classic, New York
Osmundson TH (1999) Roof gardens: history, design and construction. W. W. Norton, New York
Owens PE (2002) No teens allowed: the exclusion of adolescents from public spaces. Landsc J 21(1):156–163
Paine TD, Millar JG, Hanks LM (1995) Integrated program protects trees from eucalyptus long-horned borer. Calif Agric 49:34–37
Parkinson S (1784) A journal of a voyage to the South Seas, in his Majesty's ship, the Endeavor. Printed for Charles Dilly and James Phillips, London
Payne R (translator) (1966) Strabo Walahfrid *Hortulus*. Hunt Botanical Library, Pasadena
Penn W (1681) A short advertisement. In William Penn Papers (Instructions). (See Hough O (1895) Captain Thomas Holme, Surveyor-General of Pennsylvania and Provincial Councilor. Pe Mag Hist Bibliogr 19(1895):414–418)
Pepper PJ, McPherson EG, Simpson JR, Gardner SL, Vargas KE, Xiao Q (2007) New York City, New York Municipal Forest Resource analysis. Center for Urban Forest Research. USDA Forest Service. Pacific Southwest Forest Research Station, Davis
Pepys S (2001) The diary of Samuel Pepys. Modern Library, New York
Perry B (1989) Trees and shrubs for dry California landscapes. Land Design Publishing, Claremont
Pickett STA (1989) Space-for-time Substitution as an alternative to long-term studies. In: Likens GE (ed) Long-term studies in ecology. Springer, New York
Pinkney DH (1958) Napoleon III and the rebuilding of Paris. Princeton University Press, Princeton
Pinos DPB (1942) The Exposicion, 1812. In three New Mexico chronicles. Translated and annotated by Bailey Carroll H, Villasana Haggard J. The Quivira Society, Albuquerque
Pizzoni F (1999) The garden: a history in landscape and art. (English translation by Landry J). Rizzoli International Publications, Pittsburgh/New York
Plato (2000) Laws (trans: Jowett B). Macmillan, New York (Law 12.947)
Pliny the Elder (77 AD) Natural history. English translation by Rackham H 1938–62. Harvard University Press, Cambridge, MA
Plotkin A (2000) The urban tree book. Three River Press, New York
Polunin O, Walters M (1985) A guide to the vegetation of Britain and Europe. Cambridge University Press, Cambridge
PortPhilip Ecocentre (1998) Gardens in the sky: rooftop greening project. http://www.portphillip.vic.gov.au/ecocentre/rooftop/index.html
Powell SC (1965) Puritan village: the formation of a new England town. Doubleday, Garden City
Prescott WH (1937) History and conquest of Mexico and history of the conquest of Peru. Cooper Square Press, New York, p 2000
Price ET (1968) The central courthouse square in the American county seat. Geogr Rev 58:29–60
Profous GV (1992) Trees and urban forestry in Beijing, China. J Arboric 18(3):145–154
Raunkiaer C (1905) Types biologiques pour la geographie botanique. Bul Acad Roy Sci Denmark 5
Raunkiaer C (1934) The life forms of plants and statistical plant geography. Clarendon Press, Oxford
Report of the Royal Commission on London Parks (1928) HMSO, London
Reps JW (1969) Town planning in frontier America. Princeton University Press, Princeton
Ridington WR (1935) The Minoan-Mycenaean background of Greek athletics. University of Pennsylvania, Philadelphia
Riley AL (1998) Restoring streams in cities: a guide for planners, policymakers, and citizens. Island Press, Washington, DC

Roberts W (1763) An account of the first discovery and natural history of Florida. London (Map of St Augustine drawn by Thomas Jeffreys)
Roche P (2005) Aristophanes. New American Library, New York
Rogers DJ, Rogers C (1991) Woody ornamentals for deep south gardens. University of West Florida Press, Pensacola
Roloff A (2015) Urban tree management: for the sustainable development of green cities. Wiley, New York
Rome (1588) Archivio di Stato. Plan for Villa Lante
Rowntree RA (1984) Ecology of the urban forest-introduction to part I. Urban Ecol 8:1–11
Rowntree RA (1986) Ecology of the urban forest-introduction to part II. Urban Ecol 10:1–10
Rowntree R (1988) Ecology of the urban forest: introduction to part III. Landsc Urban Plan 15:1–10
Rybczynski W (1999) A clearing in the distance: Frederick Law Olmsted and America in the nineteenth century. Scribner, New York
Rydberg D (1998) Urban forestry in Sweden. Acta Universitatis Agriculturae Sueciae Silvestra 73
Rydberg D, Falck J (1998) Designing the urban forest of tomorrow – pre commercial thinning adapted for use in urban areas in shaping tomorrow's urban forest. Arboric J 22:147–171
Salisbury EJ (1943) The flora of bombed areas. Nature 151:462–466
Sapori A (1940) Studi di storia economic a medieval. G. C. Sansoni, Rirenze
Scharenbroch BC, Morgenroth J, Maule B (2015) Tree species suitability to bioswales and impact on the urban water budget. J Environ Qual 45(1):199–206
Schein RD (1993) Street trees: a manual for municipalities: a comprehensive but simple guide to the establishment and maintenance of shade trees on the streets of small towns and cities. Tree Works, State College
Schmid JA (1975) Urban vegetation: a review and Chicago case study, Research paper No. 101. Department of Geography, University of Chicago, Chicago
Schneider SH (1993) Scenarios of global warming. In: Karieva PM et al (eds) Biotic interactions and global change. Sinauer Associates, Sunderland, pp 9–23
Scholz H (1960) Die Veranderungen in der Ruderalflora Berlins. Ein Beitrag zur jungsten Florengeschichte Willdenowia 2:379–397
Scholz-Barth K (2001) Green roofs: storm management from the top down. Environmental Design and Construction
Schuyler D (1986) The new urban landscape: the redefinition of city form in nineteenth-century America. Johns Hopkins University Press, Baltimore
Sexby JJ (1898) The municipal parks, gardens, and open spaces of London: their history and associations. Elliot Stock, London
Sharif ME (2003) Trees occurring in Dubai. Dubai Municipality. Public Parks and Horticulture Department
Shelford VE (1932) Basic principles in the classification of communities and habitats and the use of terms. Ecology 13:105–120
Shepherd JC, Jellico G (1953) Italian gardens of the renaissance. A. Tiranti, London
Siren O (1949) Gardens of China. Ronald Press Co., New York
Sjoberg G (1965) The origin and evolution of cities. Sci Am (September, 1965)
Smith B (1943) A tree grows in Brooklyn. Harper & Brothers, New York
Smith DS (1993) An overview of greenways. In: Smith DS, Hellmund PC (eds) Ecology of greenways. University of Minnesota Press, Minneapolis
Southworth M, Ben-Joseph E (1997) Streets and the shaping of towns and cities. McGraw-Hill, New York
Spirn A (1984) The granite garden. Basic Books, New York
Spongberg SA (1990) A reunion of trees. Harvard University Press, Cambridge, MA
Stilgenbauer J, McBride JR (2010) Reconstruction of urban forests in Hamburg and Dresden after World War II. Landsc J 29(2):144–161

Stilgoe JR (1982) Common landscape of America, 1580 to 1845. Yale University Press, New Haven
Strabo (64–23 BC) Geographical sketches. English translated by H.L. Jones. 1917–28. G. P. Putnam's Sons, New York
Strand SE et al (1995) Phytoremediation of trichloroethylene from polluted aquifers using poplars. In: Proceedings of the international poplar symposium: populus biology and its implication for management and conservation, 20–25 August, 1995, Seattle, p 88
Sukopp H (1973) Großstadt als Gegenstand ökologischer Forschung. Schrift Ver Verbreitung naturwiss Kennin Wein 113:90–140
Sukoop H (2002) On the early history of urban ecology in Europe. Preslia, Praha 74:373–393
Sullivan C (2002) Garden and climate. McGraw-Hill, New York
Tabatabaie M (2005) Natural resources of Kurdistan. University of Iran, Karajh
Taha H (1996) Modeling impacts of increased urban vegetation on ozone air quality in the South Coast Air Basin. Atmos Environ 30(20):3423–3430
Tammita-Delgoda S (1995) A traveller's history of India. Interlink Books, New York
Tansley AG (1935) The use and abuse of vegetational concepts and terms. Ecology 16(3):284–307
Thacker C (1979) The history of gardens. University of California Press, Berkeley
Thaiutsa B (2000) Street trees of Bangkok. Asksorn Siam Printing, Bangkok
Theophrastus (372–287 BC) Enquiry into plants. English translation by Sir Arthur Holt. 1916. Harvard University Press. Cambridge. Two Vols
Thompson N (2004) The plight of Japanese pine. Japan File Kansai Out, May 2004
Thornthwaite CW (1931) The climates of North America according to a new classification. Geogr Rev 21:633–655
Thornthwaite CW (1948) An approach towards a rational classification of climate. Geogr Rev 38:55–94
Thornthwaite CW, Hare FK (1955) Climate classification in forestry. Unasylva 9:50–59
Tidwell J (2001) The twenty-five worst cities to live in due to pollen allergies. About Newsletter. http://allergies.about.com
Tishler WH (1989) American landacape architecture. Preservation Press, Washington, DC
Tongiorgi T (1983) Projects for botanical and other gardens: a sixteenth century manual. J Gard Hist 3:1–34
Trees for Life (2004) http://www.treesforlife.org
Trowbridge PJ, Bassuk NL (2004) Trees in the urban landscape: site assessment, design, and installation. Wiley, Hoboken
Tsuru K (ed) (1992) Elements and total concept of urban tree design. Graphic-sha, Tokyo
Turner T (1991) Toward a green strategy for London. London Planning Advisory Committee, London
UNDP (1998) Human development report 1998. United Nations Development Program. Oxford University Press, Oxford
Urban J (1989) New techniques in urban tree planting. J Arboric 15:281–284
USDA (2002) Pest Alert –Emerald Ash Borer (August 2002). http://www.na.fs.fed.us/spfo/pubs/pest_al/eab/eab.htm
U.S. Strategic Bombing Survey (1946) The effects of atomic bombs on Hiroshima and Nagasaki, U.S. strategic bombing survey. Pacific War Report No. 3. U.S. Govt Printing Office, Washington, DC
Valk H (2006) Cemeteries and ritual meals. In: Andren A, Jennbert K, Raudvere C (eds) Old Norse religion in long-term perspectives: origins, changes and interactions – an international conference in Lund, Sweden, 3–7 June 2004. Nordic Academic Press, Lund, pp 141–146
Valla JJ et al (1999) Arboles Urbanos. Edicion L.O.L.A, Buenos Aires (in Spanish)
Vervaeke P et al (2003) Phytoremediation prospects of willow stands on contaminated sediments: a field trial. Environ Pollut 126(2):275–282

Vessel MF, Wong HH (1987) Natural history of vacant lots. University of California Press, Berkeley
Violich F (1985) Urban planning in Latin America: the challenge of metropolitan growth. College of Environmental Design, Berkeley
von Fischer-Benzon RJD (1894) Altdeutsche Gartenflora. Keil und Leipzig. p 183
von Humboldt A (1805) Essai sur la geographie des plantes. Paris
Walter H, Harnickell E, Mueller-Dombois D (1975) Climate-diagram maps. Springer, New York
Warner SB (1987) To dwell is to garden: a history of Boston's community gardens. Northeastern University Press, Boston
Watkin D (1982) The English vision: the picturesque in architecture, landscape, and garden design. Harper and Row, New York
Watkins JV, Sheehan TJ (1975) Florida landscape plants. Rev edn. University Press of Florida, Gainesville
Whittle T (1997) The plant hunters. Lyons and Burford, New York
Wilkinson A (1998) The garden in Ancient Egypt. Rubicon Press, London
Williams D (2004) Seattle embarks on a dramatic experiment in restoration. High Country News, May 10, 2004
Wilson WH (1989) The city beautiful movement. Johns Hopkins University Press, Baltimore
Woods M (1996) Visions of Arcadia: European gardens from renaissance to rococo. Aurum Press, London
Wright R (1939) The story of gardening. Dover Publications, New York
Wright HE Jr (ed) (2003) Late-quaternary environments of the United States. University of Minnesota Press, Minneapolis
Wroth WW (1896) The pleasure gardens of the eighteenth century. Archon Books, London
Wu Z (2004) Personal Communication. Faculty of Forestry, Anhui Agricultural University, Hefei
Xenophon. 430–355 BC. Xenophon in seven volumes. 1968. Harvard University Press, Cambridge, MA
Yang J, McBride JR (2003) A unique technique for street tree planting in Beijing. Arboricult J 27:1–10
Yang J et al (2005) The urban forest in Beijing and its role in air pollution reduction. Urban For Urban Green 3:65–78
Yang J, Zhao L, McBride JR, Gong P (2009) Can you see green? Assessing the visibility of urban forests in cities. Landsc Urban Plan 91(2):97–104
Yoshimura A et al (1999) Modeling the spread of pine wilt disease caused by nematodes with pine sawyer as vector. Ecology 80(6):69–76
Yuen B, Hong Y (2002) User's survey. Handbook on skyrise greening in Singapore. National Parks Board and National University of Singapore, Singapore
Yule H, Cordier H (1992) The travels of Marco Polo. Dover Publications, New York
Zion RL (1995) Trees for architecture and landscape. Van Nostrand Reinhold, New York
Zucker P (1959) Town and square from the agora to the village green. Columbia University Press, New York

The manufacturer's authorised representative in the EU is Springer Nature Customer Service Centre GmbH, Europaplatz 3, 69115 Heidelberg, Germany. If you have any concerns regarding our products, please contact ProductSafety@springernature.com

Printed and bound by CPI Group (UK) Ltd, Croydon, CR0 4YY
05/10/2026
02228135-0008